Adapting Transport Policy to Climate Change

Carbon Valuation, Risk and Uncertainty

Research Report

Please cite this publication as:
OECD/ITF (2015), *Adapting Transport Policy to Climate Change: Carbon Valuation, Risk and Uncertainty*, OECD Publishing, Paris.
http://dx.doi.org/10.1787/9789282107928-en

ISBN 978-92-82-10791-1 (print)
ISBN 978-92-82-10792-8 (PDF)

Photo credits: Cover © Stefano Buttafoco, shutterstock.com

Corrigenda to OECD publications may be found on line at: *www.oecd.org/about/publishing/corrigenda.htm*.

THE INTERNATIONAL TRANSPORT FORUM

The International Transport Forum at the OECD is an intergovernmental organisation with 57 member countries. It acts as a strategic think-tank, with the objective of helping shape the transport policy agenda on a global level and ensuring that it contributes to economic growth, environmental protection, social inclusion and the preservation of human life and well-being. The International Transport Forum organises an Annual Summit of ministers along with leading representatives from industry, civil society and academia.

The International Transport Forum was created under a Declaration issued by the Council of Ministers of the ECMT (European Conference of Ministers of Transport) at its Ministerial Session in May 2006 under the legal authority of the Protocol of the ECMT, signed in Brussels on 17 October 1953, and legal instruments of the OECD.

The Members of the Forum are: Albania, Armenia, Argentina, Australia, Austria, Azerbaijan, Belarus, Belgium, Bosnia and Herzegovina, Bulgaria, Canada, Chile, China (People's Republic of), Croatia, Czech Republic, Denmark, Estonia, Finland, France, Former Yugoslav Republic of Macedonia, Georgia, Germany, Greece, Hungary, Iceland, India, Ireland, Israel, Italy, Japan, Korea, Latvia, Liechtenstein, Lithuania, Luxembourg, Malta, Mexico, Republic of Moldova, Montenegro, Morocco, the Netherlands, New Zealand, Norway, Poland, Portugal, Romania, Russian Federation, Serbia, Slovak Republic, Slovenia, Spain, Sweden, Switzerland, Turkey, Ukraine, United Kingdom and United States.

The International Transport Forum's Research Centre gathers statistics and conducts co-operative research programmes addressing all modes of transport. Its findings are widely disseminated and support policy making in member countries as well as contributing to the Annual Summit.

Further information about the International Transport Forum is available at

www.internationaltransportforum.org

Table of contents

Boxes

Figures

Tables

Abbreviations

CAPM	Capital Asset Pricing Model
CBA	Cost-benefit analysis
CE	Certainty equivalent
CEA	Cost-effectiveness analysis
CO2	Carbon dioxide
DCCEE	Australia's Department of Climate Change and Energy Efficiency
DECC	UK's Department of Energy and Climate Change
Defra	UK's Department for Environment, Food and Rural Affairs
ETS	Emission Trading Scheme
EU	European Union
GDP	Gross Domestic Product
GHG	Greenhouse gas
IAMs	Integrated Impacts Assessment Models
IPCC	Intergovernmental Panel on Climate Change
PPP	Purchasing power parity
SCC	Social cost of carbon (typically estimates from the damage costs approach)
SOC	Social opportunity cost of capital
SRTP	Social Rate of Time Preference
USG	United States Government
WACC	Weighted average cost of capital
WTP	Willingness to pay

Executive summary

Transport accounts for nearly a quarter of carbon dioxide emissions from fuel combustion. The way these emissions are considered in economic appraisals of transport policies and investments in the transport sector has a significant impact on climate policy and trade-offs made between mitigation of climate change and other policy objectives. Inappropriate valuation of carbon emissions will affect the level of mitigation achieved and is likely to undermine social welfare through an inappropriate allocation of resources. There are three inter-related issues around incorporation of climate change effects in transport appraisals. They are the valuation of carbon dioxide, the treatment of uncertainty and the approach used to discount future costs and benefits.

The valuation of carbon dioxide emissions (or carbon value for brevity) is subject to high uncertainty due to the complex feedback effects between emissions, mitigation, adaptation and carbon dioxide concentration in the atmosphere. While climate models to estimate likely impacts on the climate and resulting damage costs have greatly improved, these models are still subject to limitations. In theory, alternative approaches to deriving carbon values can substitute for damage cost estimates. Abatement costs and carbon market prices have been employed by many jurisdictions as substitutes for damage cost estimates. Under certain conditions these approaches should yield the same estimate as climate models, notably if mitigation policies are set at the optimal level and if no distortion exists in emissions trading markets. In reality these three approaches yield different results. Thus far, there is no consensus internationally on the approach to use to value carbon dioxide emissions.

To account for long-term uncertain climate effects in transport appraisal, it is necessary to distinguish between risk, i.e. uncertainty that is characterised by an objective probability distribution, and unquantifiable uncertainty or "Knightian" uncertainty (after Knight, 1921). Risk can be dealt with by incorporating the probability distribution of climate events in cost-benefit and related analyses. For Knightian uncertainty, other techniques could be used to supplement or inform the cost-benefit analysis (CBA). Presenting a separate uncertainty analysis to decision-makers will assist in making better policy and investment decisions.

CBA requires the conversion of future costs and benefits into present values through discounting. Ramsey (1928) proposed a social rate of time preference formula (i.e. the Ramsey formula) to do this. This formula reflects the impact of savings and investment on consumption and the time preference individuals have for consumption today over the same level of consumption at a later date. To account for consumption risk, the standard Ramsey formula can be extended by subtracting a precautionary term that reflects the tendency for individuals to invest more for the future by reducing the discount rate. The resulting formula is often referred to as the extended Ramsey formula. Projects are also usually subject to systemic risk linked to uncertain macroeconomic conditions. A systemic risk premium can be added to the extended Ramsey formula to account for this. The risk premium is likely to increase with uncertainty. Therefore, when the precautionary effect is combined with the systemic project risk effect, the discount rate may increase or decrease over time, depending on the relative weights of the two effects.

In recent literature, models have been developed to examine how uncertainty affects the discount rate by applying a subjective probability distribution over objective probability distributions. Early results indicate that a decision-maker who is more averse to uncertainty will have a lower discount rate. As this area of research develops, practical steps to account for Knightian uncertainty may become possible.

To conclude, risk and uncertainty are key challenges for incorporating climate effects in transport appraisal. They affect the assessment of the value to place on carbon emissions as well as the quantification of climate impacts. Risk and uncertainty also affect the choice of discount rate, which in turn also affects the valuation of carbon through the damage cost modelling process. This report presents techniques to deal with risk in transport appraisals, such as risk-adjusted discount rate. More research is still required to develop a robust framework for addressing uncertainty, e.g. in setting the discount rate schedule. Below is a list of key messages for decision-makers.

Key messages for decision-makers

- **Uncertainty is different from risk.** Risk refers to cases where it is possible to estimate the likelihood of occurrence for each possible state. (*Known unknowns* in popular expression). Uncertainty on the other hand, refers to situations where it is not possible to assign a robust probability distribution around the possible outcomes (*unknown unknowns* or in technical terms Knightian uncertainty).
- **Climate effects are subject to uncertainty.** The unknowns around climate effects largely fall under the category of uncertainty. Some climate change effects are largely unknowable due to modelling limitations. It is also unknown how the economy and society might respond to climate effects in the future.
- **There are techniques to deal with risk.** Risk can be addressed with a number of techniques. These include risk-adjusted discount rates for incorporation in cost-benefit analysis and carrying out scenario analysis, sensitivity testing and simulation exercises to produce a likely range of outcomes from cost-benefit analysis.
- **There is currently no robust method to treat Knightian uncertainty.** However, models are being developed in this area and there may be a practical approach to account for uncertainty in the future as research matures. In the interim, decision-makers should be provided with a separate uncertainty analysis to highlight the key parameters that are subject to uncertainty and to explain how uncertainty could alter overall costs and benefits of policy or investment proposals.
- **Risk, uncertainty and discount rate all affect carbon value**. Risk and uncertainty affect how people respond to climate effects by changing their consumption patterns and hence affect the implicit discount rate. These changes vary over time as the level of risk and uncertainty increases. This means the discount rate may vary over time. The choice of discount rate schedule will affect the valuation of carbon through the modelling process. It also has significant impact on the discounted carbon value used in a CBA. Even with a discount rate that would be considered moderate in most investment contexts (e.g. 3-5% per year) the discounted carbon value will be significantly smaller than the undiscounted carbon value.

Chapter 1

Summary and recommendations

This chapter summarises the three key aspects of the assessment of climate change effects in transport appraisal considered in this report: the valuation of carbon; the treatment of risk and uncertainty; and methods for discounting long-term effects.

Valuation of carbon

This report examines four types of estimates for the cost of carbon commonly used to inform policy and investment appraisals. They are damage costs, two variants of abatement costs and the carbon market price. These estimates are established for different purposes. Only the damage cost estimate attempts to measure the marginal cost of carbon emissions.

The price of carbon on current emissions trading markets is strongly influenced by decisions on the total number of emission allowances to be issued and by the exemptions accorded to specific sectors of the economy. As limits on emissions allowances have tended to be relaxed over time and large parts of the economy are not included in the trading systems in operation so far, current prices are unlikely to reflect efficient abatement cost levels and are extremely unlikely to give a useful indication of the social cost of CO_2 emissions. Therefore, current market prices are not suitable for use as a direct indicator of the cost of carbon in transport CBA.

The value of carbon derived from the abatement cost approach varies with the policy options selected in the emissions mitigation strategy and with the CO_2 emissions target chosen. It will reflect the true cost of CO_2 emissions only if the emission target is set at the economically optimal level. If countries believe their emission target is set at the right level, then cost of abatement can be used in CBA without creating internal inconsistencies, especially if CBA is performed from a national and not a global perspective.

Theoretically, the damage cost approach is the most appropriate approach for assessing the climate effects of CO_2 emissions. The approach reflects mitigation options and adaptation potential in calculating costs. A range of Integrated Impact Assessment Models (IAMs) have been developed by scientists and modellers to calculate damage costs. The resulting estimate is usually referred as the social cost of carbon (SCC). The SCC intends to be a comprehensive measure of a wide range of climate effects covering environmental, social, economic, health and ecosystem effects. In practice, estimating the SCC is very difficult and modelling work is still subject to a number of limitations including inconsistent assumptions between models, potentially inappropriate assumptions and omitted effects.

At this stage, there is no consensus as to which carbon value estimates should be adopted globally. Different jurisdictions adopt different values and different approaches to estimating the cost of carbon. In some jurisdictions different values for carbon are used in appraisal from sector to sector.

There is, however, consensus that carbon values are not constant over time as the impact of the emission of an extra tonne of CO_2 will vary depending on the current concentration of CO_2 in the atmosphere. As the concentration of CO_2 in the atmosphere is expected to rise with emission level, a number of jurisdictions apply different carbon values to different time periods in appraisal. This report shows that the choice of discount rate schedule can have significant impact on the discounted carbon value used in a CBA. Even with a moderate discount rate (e.g. 3-5% per year), the discounted carbon value will be much lower than the undiscounted value.

Since future carbon values are likely to be subject to high uncertainty and change over time, it is perhaps more important to ensure CBA consider these uncertainties than to determine a point estimate for the value of carbon.

Recommendations on the valuation of carbon

- A common carbon value should be assigned for national investment and mitigation policy appraisal in all sectors within the same jurisdiction.
- There is merit in countries working together to develop a set of principles on how carbon values used in assessment should vary over time in real terms.

Risk and uncertainty

Treatment of climate change risk

As distinguished by Knight (1921), measurable uncertainty refers to risk that can be approximated by a statistical distribution of possible outcomes whereas uncertainty refers to circumstances where statistical quantification is not possible. This distinction has important implications because it means that methodologies designed to reflect risk in CBA do not address the issues associated with uncertainty.

This report looked at possible tools to supplement CBA in face of climate change risk. These include adoption of risk-adjusted discount rates and conduct of separate impact assessments, cost-effectiveness analysis, sensitivity testing or scenario testing; and the use of real-option approaches to improve the development and selection of projects. While CBA continues to play a key role to inform the value for money of policy or investment decisions, these supplementary analyses can improve the richness and robustness of the results of a conventional CBA.

Treatment of climate change related uncertainty

Climate change effects are subject to two types of uncertainty – scientific uncertainty resulting from incomplete knowledge about the climate system and socio-economic uncertainty due to unknowns about how societies and economies will function in the future and how they will respond to climate effects.

To ensure the quality of policy and investment decisions, decision-makers need to be informed about how uncertainties (e.g. around future demand, economic conditions and climate impacts) affect the estimated costs and benefits of an intervention or investment. This can be done by carrying out an uncertainty assessment, which will be similar to sensitivity testing or scenario analysis but rather than testing a distribution of statistically likely outcomes around a mean will assess project or investment outcomes under a number of different uncertain future states that cannot be assigned a statistical likelihood. The assessment will provide decision-makers with explicit information about the uncertainties involved and how they impact on the overall cost and benefit positions. In the absence of a statistical technique to account for uncertainty, a separate uncertainty assessment will be of value to decision-makers.

In practice, this would mean providing a likely range of results after considering risk and another wider range of results that consider the impacts of uncertainty. The latter will need to be supported by descriptions of the sources of uncertainty, its determinants and potential impacts.

Recommendations on the treatment of risk and uncertainty

- CBAs should be supplemented with information on long-term impacts that are subject to a high level of uncertainty.

- Risk should be factored into CBA by using appropriate tools such as adoption of risk-adjusted discount rate or use of sensitivity analysis or scenario testing.
- A separate uncertainty assessment should be carried out to better inform decision-makers on the potential impacts of uncertainty on the costs and benefits of an intervention or investment decision.

Discounting under risk and uncertainty

Discounting is an integral part of CBA for policy or project appraisals with costs and benefits that spread over a number of years. The choice of discount rate has a significant impact on assessment outcomes. There is no consensus on what discount rate to use. The uncertainty involved in estimating the climate effects, which affect generations further into the future most, complicates selection of an appropriate discount rate schedule to use in CBA.

The Ramsey formula is usually used as a basis for determining intergenerational discount rates. The Ramsey formula can be extended by subtracting a precautionary term to account for consumption growth risk (extended Ramsey formula). To account for systemic project risk, a risk premium term can then be added (systemic risk-adjusted Ramsey formula).

Uncertainty around interest rates and/or components of the social discount rate, such as consumption growth or expected project benefits, can both affect the choice of the discount rate used in CBA. In the absence of project risk, the discount rate would be close to the risk-free rate. In this situation, taking account of uncertainty (by introducing a precautionary factor in the extended Ramsey formula) reduces the discount rate. The adjustment is larger the greater the uncertainty and this can justify adoption of a declining risk-free discount rate.

If there are project risks, the risk premium component of the discount rate is likely to increase with uncertainty. The overall effect of uncertainty is thus ambiguous. If the effect of uncertainty on the risk-free rate is less than the effect on the systemic risk premium, the discount rate (in the systemic risk-adjusted Ramsey formula) may increase over time.

Thus far, common methodologies to consider uncertainty in the discount rate can only capture risk. More recent literature has developed models to capture part of the Knightian uncertainty in the discount rate by applying a subjective probability distribution over the objective probability distribution. As research matures, practical steps to establish a discount rate that can partly capture some aspects of uncertainty should become available.

Recommendations around discounting under risk and uncertainty

- As the debate around which discount rate method to use and what parameter values should be used within the chosen method is unlikely to be resolved in the near future, one way to reflect 'Knightian uncertainty' is to carry out a sensitivity analysis of CBA results using a high and a low (but constant) discount rate.
- On-going work on reflecting uncertainty in discounting through statistical techniques shows promise and its value in informing long-term policy and investment decisions should be kept under review.

Concluding remarks and next steps

Risk and uncertainty are key challenges for incorporating climate effects in transport appraisal. They affect the assessment of the value to place on carbon emissions and the quantification of climate impacts. Risk and uncertainty also affect the choice of discount rate, which in turn affects the valuation of carbon through the damage cost modelling process. This report presents techniques to deal with risk in transport appraisals, such as risk-adjusted discount rate. More research is still required to develop a robust framework for addressing uncertainty, e.g. in setting the discount rate schedule.

This report can be extended by carrying out the following additional analysis:

- Identify international best practices around life-cycle assessment in transport appraisal and understand the impact of such practices on the estimated costs and benefits of transport interventions.
- Identify other international CBA practices that affect the valuation of carbon and the estimated costs and benefits of proposals.
- Examine the desirability of harmonising key CBA practices (e.g. discount rate, carbon value, evaluation period and procedure for treating residual value) and identify the benefits from doing so.

References

Knight, F. (1921), *Risk, uncertainty and profit*, Boston, MA: Hart, Schaffner & Marx; Hougton Mifflin Company.

Ramsey, Frank P. (1928), "A mathematical theory of saving", *Economic Journal*, Vol. 38(152), pp. 543-559.

Chapter 2

Challenges for including climate change effects in transport appraisal

This chapter lays the foundation for the remainder of the report. It first provides background on the nature of transport appraisal, climate change effects and the challenges for including these effects in appraisals. The remainder of the chapter defines the scope and structure of the report.

Background

Human induced warming of the climate is a major policy issue, shared by all countries. It is of particular importance to transport decision-makers because transport produces around 25% of carbon dioxide emissions from fuel combustion. Transport policies and investment decisions can influence transport activities, which in turn affect vehicle emissions and the associated climate change impacts. Current appraisal frameworks have several limitations when it comes to assessing climate change effects.

Conventional transport appraisal is based largely on the cost-benefit analysis (CBA) framework, which considers the trade-offs between socio-economic benefits and socio-economic costs. CBAs typically require monetisation of a range of benefits and costs to enable various effects to be compared using a common metric. Climate change effects are subject to a high level of uncertainty. In particular, it is difficult to determine the benefits from mitigating catastrophic climate impacts that have a low probability of occurrence but high potential impacts.

There are three key challenges in considering the effects of carbon dioxide (CO_2) emissions (referred to as carbon impacts for brevity) in transport appraisal in the CBA context. The first relates to estimating the physical relationships and impacts (including effects of CO_2 emissions on atmospheric concentration, and the effects of concentration on temperatures). The second relates to the valuation of carbon dioxide (including any economic effects resulting from climate effects). The third concerns the treatment of long-term uncertain impacts in project appraisals.

Many transport interventions and infrastructure investments can result in costs and/or benefits that span decades. The cumulative nature of greenhouse gases (GHGs) such as carbon dioxide means that "the climatic impact of any given increase in the [GHG] stock in a given year is not confined to that year and may persist over many years" (Rhys, 2011). As GHG concentrations in the atmosphere rise, incremental emissions are likely to produce larger effects. This means future emissions are expected to produce larger incremental damages, implying they should be assigned a higher value or cost (in real terms). The level of uncertainty related to the timing and outcomes of some climate change impacts means it is difficult to accurately determine the resulting economic impacts and the value to place on reducing carbon dioxide emissions.

There are three interrelated issues regarding the challenge of putting a value on carbon:

- While there has been significant development in modelling the impacts of climate change on the environment, people and the economy, countries that use the same approach often arrive at different estimates due to differences in data, models and/or assumptions used.
- Several different approaches to valuing carbon have been used for climate policy appraisal. Estimates of the marginal cost of damage from climate change provide the basis for estimating the marginal *benefit* of abating CO_2 emissions. Estimates of the cost of reducing CO_2 emissions by a tonne (now or in the future with improved technology) provide a measure of the marginal *cost* of abatement. These estimates measure different things and each approach has its own limitations.
- There is evidence of different values being used by different sectors even within the same jurisdiction.

Inappropriate and inconsistent assessment of the value of carbon will undermine the efficiency of policy responses and increase costs.

Uncertainty in estimating elements of a CBA, especially when the element accounts for a significant share of costs or benefits, could make the outcome questionable and potentially unreliable. Accounting for low probability but potentially extremely severe climate impacts in CBA is not an easy task. Views are divided as to whether and how such damages should be taken into consideration. Inclusion of this type of impact in the assessment framework may change the outcome of the analysis, but again, views are divided on how significant the difference is likely to be.

Most fundamentally, under a conventional CBA approach with a constant discount rate, a large proportion of the benefits or costs will be discounted away by the end of an assessment period of 40 to 60 years. The discount rate, which applies to future costs and benefits to derive their present values, plays an important role and can have significant influence on the outcome of policy analysis. For policies designed to address very long term effects such as climate change the influence of the discount rate is very large. Choosing appropriate discount rates is critical to the value of CBA in these circumstances.

Research questions

These challenges lead to three sets of important questions for the assessment of policies for more sustainable transport:

- **Valuation of carbon**: What is the correct approach to value the social cost of carbon emissions? How should the carbon value evolve over time? Should the carbon value be the same across different sectors within a jurisdiction? Should the carbon value be the same across jurisdictions?
- **Climate change related uncertainty**: How does uncertainty affect the results of traditional cost-benefit analysis? Does the scale of uncertainty associated with some climate change effects require special treatment?
- **Discounting long-term effects**: Should the discount rate be constant over time? If a declining discount rate schedule should be used in project appraisal, how should the schedule be determined?

Drawing from international literature, results from a preliminary OECD survey of monetary carbon values in selected member countries and discussions at the ITF working group meetings of 12-13 December 2013, 27 February 2014 and 10 October 2014, this report discusses these three issues in turn.

Scope of the report

This report covers only the key research questions outlined in section 1.2. The working group has identified other research areas that are also important when incorporating climate effects in transport appraisal. Due to time constraints, the topics listed below are not included in this report and need to be addressed separately.

- **Challenges around quantification of CO_2 emissions** – Due to imprecise estimates of factors such as emission intensity, vehicle occupancy rates and modal splits, estimates of CO_2 emissions for a given jurisdiction can be highly variable. These challenges have important implications because errors and subsequent improvements can make a significant difference in the cost and benefit estimates. Since the present report focuses on valuation and long-term issue, this subject is excluded from the discussion.

- **Embedded carbon** – Greenhouse gas emissions from transport sector can result directly from the use of fossil fuels and indirectly from the use of materials for the construction and maintenance of transport infrastructure (e.g. production of steel and aggregates may require burning fossil fuels) or from the disposal or recycling of transport wastes (e.g. tyres and scrap metals). For some projects, a CBA should take a whole-of-life perspective to understand the impact of transport decisions on carbon emissions. Issues for CBA in relation to life-cycle assessment also include evaluation horizon and treatment of residual values, which are not discussed in this report. A survey of international practices and framework for carrying out a life-cycle analysis could usefully be developed as a separate exercise.
- **Other health and environmental effects due to emissions** – Transport emissions have significant health and environmental effects. Presently, the Environment Directorate of the OECD is carrying research on these issues under the "Cost of Inaction and Resource Scarcity; Consequences for Long-Term Economic Growth". Amongst other things, the research will examine health impacts from air pollution, quantifications of macroeconomic costs of climate effects through climate modelling; the consequences of reduced availability or quality of land and assessment of the consequences of the loss of biodiversity and ecosystems. In view of the broader work programme within OECD Environment Directorate, the current report has excluded these issues.
- **Internal and external cost of carbon emissions** – National cost-benefit analysis should cover both internal and external costs. However, in the case of climate effects the situation is complicated by the presence of emissions trading schemes (ETS). Sectors included in the ETS will need to either reduce emissions or purchase additional emissions units in order to meet the emission allowance. When emission units are purchased from the ETS markets, there will be no net change in the emission level. In those situations, there may not be a need to place a value on carbon for the purposes of estimating the external cost of carbon emissions in project appraisal as there is no change in emissions level. On the other hand, if the sectors choose to reduce emissions instead, the benefits of the CO_2 reduction may be higher than the carbon market price; to date carbon market prices do not generally reflect damage cost of emissions. In those situations, it may still be necessary to place a value on carbon to understand such benefits. The internalisation impacts of ETS on carbon value will need to be explored separately in detail.

Structure of report

The structure of the remainder of this report is as follows:

- Chapter 3: Estimating the social cost of CO_2 emissions.
- Chapter 4: Uncertainty and transport appraisal of climate change effects.
- Chapter 5: Discounting long-term effects of climate change for transport.

Reference

Rhys, J. (2011), "Cumulative carbon emissions and climate change: Has the economics of climate policies lost contact with the physics?", The Oxford Institute for Energy Studies, University of Oxford.

Chapter 3

Estimating the social cost of CO_2 emissions

This chapter considers the alternative approaches to estimating the social cost of CO_2 emissions that can be applied in cost benefit analysis (CBA). The first sections describe the analytical approaches, and how values can vary through time. The final section describes how selected OECD countries value carbon in their jurisdiction.

Approaches to valuation of carbon emissions

Valuation of carbon aims to translate the implicit value of a given amount (e.g. a tonne) of CO_2 emitted into a monetary value. There are four types of estimate for the cost of carbon:

- damage cost estimate
- abatement cost estimate based on emission target
- abatement cost estimate based on current mitigation policies
- market price of carbon.

An arbitrary value can also be assigned. The next four sections discuss each analytical approach.

Damage cost estimates

Overview

The present value of the stream of future damages associated with an incremental increase (by convention, one metric tonne) in carbon dioxide (CO_2) emissions in a particular year is commonly referred to as the Social Cost of Carbon (SCC). Theoretically, the damage cost approach is the most appropriate approach for assessing the economic effects of CO_2 emissions. A range of Integrated Assessment Models (IAMs) have been developed by scientists and modellers to assess mitigation and distributional impacts and to calculate damage costs. IAMs combine the information available on the natural mechanisms behind climate change with estimates of the economic impacts in a single modelling framework, in order to estimate the physical impacts and the monetised benefits and costs. Most IAMs incorporate adaptation potential and the effects of mitigation options when estimating the damage costs.

The SCC is intended to be a comprehensive measure of climate change damage. It typically includes changes in net agricultural productivity, energy demand, human health, property damage from increased flood risk and changes in the value of ecosystem services. In practice, estimating the SCC is very difficult and modelling work is still subject to a number of limitations including inconsistent assumptions between models, potentially inappropriate assumptions and omitted effects.

To estimate the total or marginal damage cost of climate change, it is necessary to first estimate the physical impacts of climate change under given climate change scenarios including low-probability high-impact scenarios. This includes estimating how CO_2 emissions contribute to CO_2 concentration in the atmosphere, global surface temperature and the associated climate change effects. The estimated impacts are then translated into monetary terms. Estimations of the physical impacts and the monetary value are complex tasks because climate change impacts include both market and non-market goods, covering health, environment and wider social aspects, and many of the impacts may only occur in the distant future.

For market goods (e.g. agriculture), impacts may be directly estimated based on market data. For non-market goods (e.g. health), indirect measures using revealed preference approaches and stated preference approaches are required. Revealed preference approaches measure the value indirectly on the basis of market prices for surrogate products or services (e.g. house price to measure the value of environment amenity). Stated preference approaches ask the public directly about their willingness to pay (WTP) for an improvement in their health, traffic safety and environmental quality, etc. Irrespective of whether goods are traded in marks or not, obtaining future values is a challenge for understanding the SCC.

In a review of SCC estimates (211 estimates from 47 studies) using different IAMs, Tol (2008) (cited in Mandell, 2013) found a wide distribution of carbon value estimates,[1] from –EUR 1 to EUR 451 per tonne of CO_2. A number of factors affect the value of carbon (Figure 3.1). Many of these factors are self-explanatory but two of them are worth mentioning here. Since climate change has global impacts, aggregating damages require summing up effects from different localities (or countries). The way the aggregation is completed (e.g. output weighted and equity weighted) can have significant impacts on the marginal aggregate damage cost. For estimating future SCC in present value terms, the choice of discount rate is also an important factor. This aspect will be discussed in Chapter 4.

Figure 3.1. **Factors affecting the social cost of carbon**

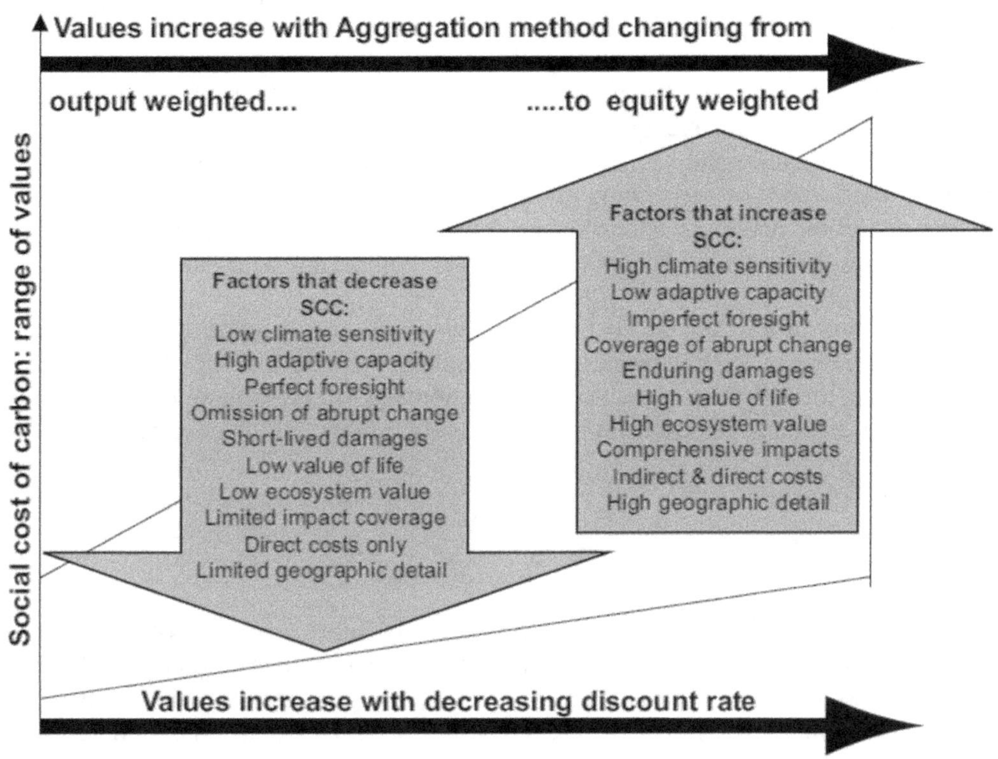

Source: Adapted from IPCC (2007).

Limitations and issues

There are several challenges in measuring marginal damages from CO_2 emissions.

The first relates to the level of uncertainty associated with climate change and its impacts. While IAMs try to count all the damages to be expected from climate change, based on the best available scientific knowledge, it is extremely difficult to specify climate impacts over a long time scale. Further, many items remain difficult to quantify (e.g. agriculture) and/or monetise (e.g. loss in biodiversity). There could be other unexpected future damage (such as the potential for "catastrophic" impacts) caused by global warming that are at best imperfectly captured in some models, and not at all in others. Therefore, many of the SCC estimates are considered as low-end estimates.

The second challenge relates to the equity concerns when aggregating impacts across very different countries. The two alternative approaches – output versus equity weighting – each have difficulties.

Under output weighting, low income countries are given little weight, even though they are more vulnerable to climate change (Kuik et al., 2008 and Anhoff and Tol, 2010). Therefore, using output weighting will result in a low estimated marginal aggregated damage costs.

Alternatively, an equity weighting can be applied during the aggregation process by first aggregating regional welfare losses, then applying monetary values (Anhoff and Tol, 2010). The equity weighting approach confronts other equity issues. Climate damage is likely to be worse in low income countries, which are likely to have lower ability and, hence, willingness to pay (WTP) for the reduction of mortality risk (e.g. in estimating Value of Statistical Life) (OECD, 2012). In addition, there may be other factors (such as risk level and attitude towards risk) that affect individuals' WTP. This raises the ethical question of whether human life should be given the same value and treated equally regardless of country of origin, or whether it should be different to reflect income and other differences.

The use of equity weighting also means climate change mitigation is valued lower in the rich countries in the model than an average person would be prepared to pay in those countries. This inconsistency has importance policy implications. Researchers attempt to overcome this by assuming different global welfare function, depending on whether the model assumes the national decision maker is altruistic towards other countries and whether decision maker compensates damages done abroad. Modelling suggests results vary widely between scenarios (Anhoff and Tol, 2010).

The third limitation of the damage cost approach relates to the uncertainty with the key parameters or assumptions used in the modelling process. These include the extent of the climate change effect, the discount rate schedule, the structure of the damage function and information on mitigation and adaptation measures. This means the damage cost estimates are usually subject to high uncertainty.

Abatement cost estimates

Overview

Abatement cost estimates are also commonly used to inform carbon policy appraisals. These estimate the marginal cost of CO_2 emission reductions, rather than estimating the damage cost. Abatement costs do not represent the social cost of carbon, except under the condition that the abatement strategy is set at the optimal level, in which case the two approaches produce the same result.

There are two common approaches to determining the cost of abatement:

- estimates based on the cost of measures required to achieve a specific emission target (e.g. a reduction by 20% by 2020)
- estimates based on the costs of current mitigation policies.

Estimating the cost of abatement involves deciding on the intended action(s) to reduce CO_2 emission and estimating how much emissions and costs will change over time with and without the policy. These estimates are then combined to establish the marginal cost of abatement per tonne of CO_2 reduced. This approach enables comparison of the cost-effectiveness of different policy options.

The cost of abatement can be measured in terms of the total resource cost to the economy or the fiscal cost to government. The fiscal cost measure is only useful for comparing the fiscal implications of different policy options. The resource cost approach is more appropriate for understanding total abatement costs.

To assist decision-making, carbon prices (e.g. under an emission trading scheme or ETS) are sometimes used as a benchmark for the cost of carbon. If the cost of abatement is lower than this benchmark it indicates that the policy may be a cost-effective measure to reduce carbon pollution (DCCEE, 2011).

Limitations

The abatement cost approach is a useful way to identify cost-effective policy measures to reduce emissions. However, it has limitations.

First, the cost of abatement represents the cost of mitigation and therefore does not represent the potential benefit from mitigation or the potential costs of inaction. Hence, using the cost of abatement in CBA will not represent the true damage cost from climate change if the emission targets are not set at an economically optimal level. This limitation may be overcome by estimating the abatement costs using an economic model that equalises carbon values across all sectors while achieving the emission targets.

Secondly, the cost of abatement varies with policy options (e.g. the use of ineffective policies such as mandating the use of biofuels will result in extremely high cost per tonne of CO_2 abated) and the CO_2 reduction target chosen.

Market price of carbon

A preliminary OECD survey indicates some countries use a carbon market price to assign a value to carbon in transport appraisals. A global emission trading scheme covering emissions from all economic sectors and with a cap on emissions that is set at an optimal level derived from climate change models, would be the most cost-effective way to mitigate climate change. Under these conditions the price at which carbon is traded would faithfully track the value of carbon. This would be the value to use in investment and policy appraisals.

Today's carbon markets do not meet these conditions. They are not global, many sectors are excluded from local and regional trading schemes and caps on emissions are not set in relation to damage cost estimates but on the basis of political acceptability. Caps are frequently revised, altering prices. The prices that result carry limited information relevant to establishing a value of carbon for project of policy appraisal (Figure 3.2).

Figure 3.2. **Carbon market prices for selected Emissions Trading Schemes**

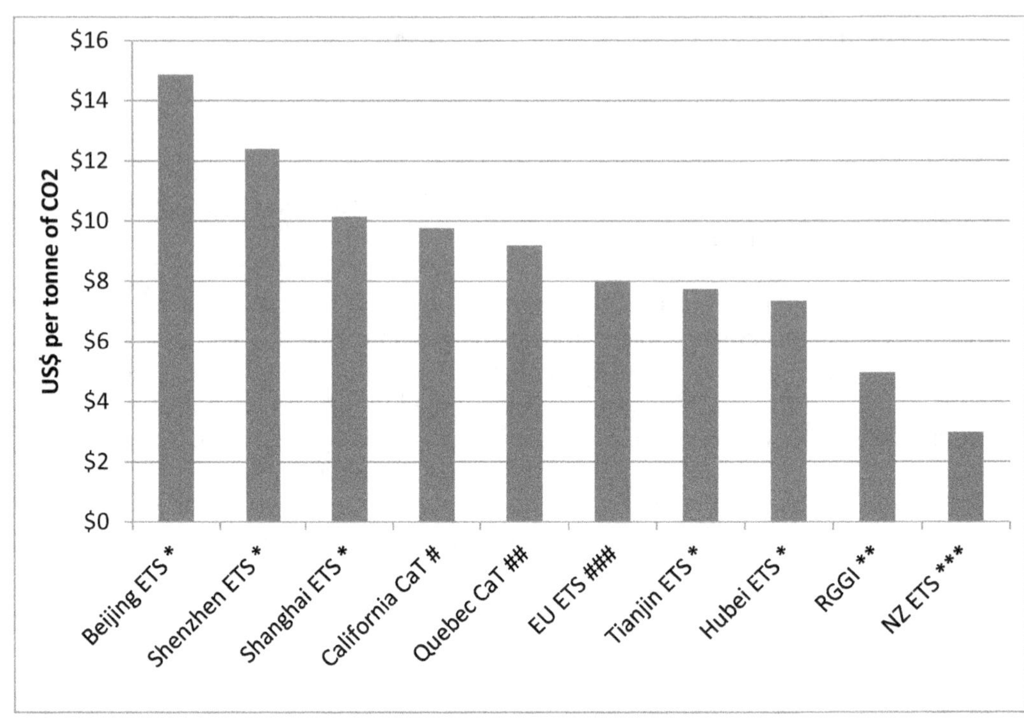

Notes: All prices have been converted to USD using Purchasing Power Parity data sourced from OECD statistics.
* Prices for 17 October 2014. Source: www.chinacarbon.net.cn.
** Price for 16 October 2014. Source: https://rggi-coats.org/eats/rggi/ .
*** Price for 17 October 2014. Source: http://www.carbonforestservices.co.nz/carbon-prices.html.
# Price for 15 October 2014. Source: http://calcarbondash.org/
Price from 26 August 2014. Source : Gouvernement du Québec (2014).
Average price in October 2014.
Source: http://www.reuters.com/article/2014/10/23/us-eu-ets-autos-idUSKCN0IC14520141023.

More broadly, carbon pricing can be an effective way to reduce emissions. Pricing emissions through a carbon tax on fuel, for example, will reduce fuel consumption and associated CO_2 emissions and stimulate development of more fuel efficient technologies. The response may not be sufficient to achieve optimal mitigation of emissions (because of market imperfections that can dilute the effect of carbon pricing) but carbon taxes are one of the most effective tools available for implementing climate policy (OECD, 2013a). The level at which the carbon tax should set raises exactly the same issues as the valuation of carbon in general. It needs to be related to damage costs or to efficient abatement costs or the clearing price in an efficient trading system. Transport fuels were taxed long before climate change became a policy issue today and are taxed for a range of purposes that differ from jurisdiction to jurisdiction. Fuels attract taxes for revenue raising purposes because fuel consumption is relatively inelastic and taxing it results in less distortion in the allocation of resources than, for example, taxing labour. In some jurisdictions fuel tax revenues are partly earmarked to fund investment in roads. Because fuel is taxed for a range of purposes, the overall level of tax on petrol or diesel is often far above the level of a carbon tax based on damage cost estimates.

Figure 3.3 summarises the wide range of prices derived from carbon pricing schemes in place in 2014. Prices in emissions trading schemes tend to be the lowest, influenced as they are by arbitrary decisions on the total number of emission allowances issued and exemptions for certain industry sectors.

Most traded carbon prices (between USD 2 and USD 14 – Figure 3.2) are much lower than estimates of the cost of abatement or the social cost of carbon generally used in transport appraisals. Carbon tax prices that are derived arbitrarily from existing taxes on fuel tend to be much higher than estimates of the cost of abatement or the social cost of carbon generally used in transport appraisals.

Figure 3.3. **Prices from carbon pricing schemes in 2014**

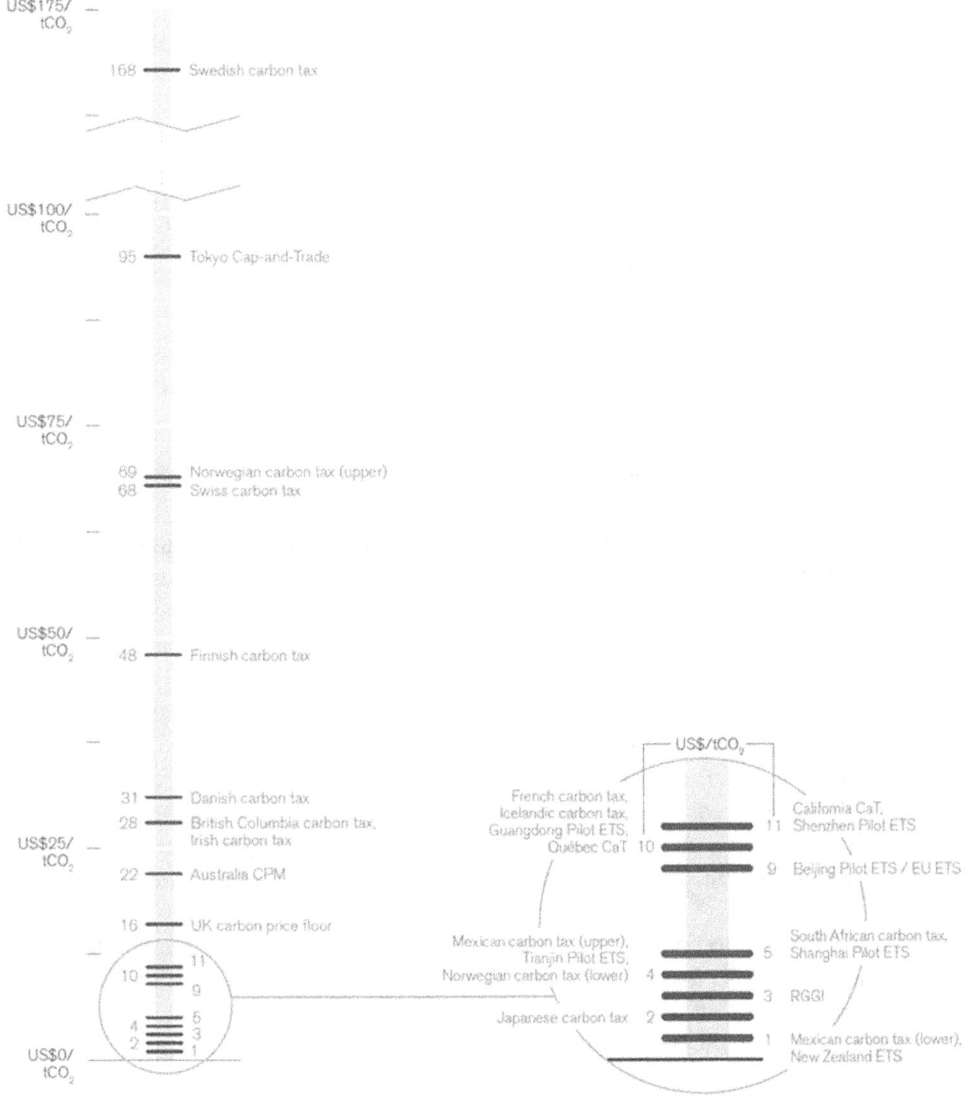

Note: The term "Pricing schemes" used in the World Bank report refers to a range of instruments including carbon taxes, emissions trading schemes and crediting mechanisms. CPM is the Carbon Pricing Mechanism, in place 2012-2014; RGGI is the Regional Greenhouse Gas Initiative; CaT is Cap and Trade.

Source: Adapted from World Bank (2014)

Another issue with using carbon pricing as a proxy carbon value is that prices can vary significantly between countries and over time. These variations have important implications for policy recommendations.

Relationships between Social Cost of Carbon, abatement cost and carbon prices

As noted, the cost of abatement is sometimes compared with a carbon tax or traded carbon prices to inform decision-makers as to whether a mitigation policy is worthwhile (e.g. compared to simply buying more carbon units in the ETS market).

Under restrictive assumptions, the cost of abatement could be the same as the Social Cost of Carbon (SCC) if the emissions target were set at the economically optimal level such that the target sufficiently reflects the damage estimation of climate change (Defra, 2007). This is, unfortunately, often not the case due to political consideration of the cost of mitigation and their impact on short-term economic prospects.

In theory, only estimates established using the damage cost approach (i.e. the SCC estimates) are appropriate for use in CBA. However, it is acknowledged that research is needed to improve on current SCC estimates. Moreover CBA is often undertaken from a national perspective while SCC is about damages assessed at the global level. In such a context, abatement costs reflecting the economic burden for a given country to meet its emission target may be a more appropriate basis for valuing carbon than a global SCC.

How should carbon value evolve over time?

Carbon value in nominal and real terms

It is important to differentiate nominal and real values when assessing climate change effects because both estimates tend to increase over time. Nominal estimates refer to estimates that are expressed in current prices (i.e. during the year when the costs or benefits occur). Real estimates refer to those that are expressed in constant price (i.e. at a chosen base-year price level which typically refers as "prices").

One of the basic requirements of CBAs is that all monetary estimates should be expressed in the same price level in order to allow consistent comparison between benefits and costs. The variations in the time series estimates of SCC in nominal terms include both the inflationary effects and the cumulative effects of climate change whereas the variations in the counterparts in real terms exclude the inflationary effects. For CBA, it is the latter that is relevant. Therefore, SCC estimates established in an earlier year (e.g. 2010 prices) need to be adjusted to the price year of the benefit and cost items (e.g. 2014 prices).

There are different ways to adjust the carbon value in real terms. If the carbon value was established based on the results of IAMs, the inputs used in the monetisation process can be updated to reflect changes in monetary values of various components. This process can be resource-demanding and therefore is not carried out on an annual basis.

A more common approach to price level adjustment is to apply an inflation adjustment factors (e.g. Consumer Price Index or GDP deflator) to a time profile of the SCC reported in real terms. For example, the SCC estimates in USG (2013) are expressed in 2007 USD. Adjusting the price level of real SCC estimates in a consistent manner should maintain the relativity between estimates in real terms, with only the price level changed.

Future carbon value

Climate damage is expected to increase over time as CO_2 concentration increases. Further, the damage functions typically used in IAMs imply that climate damage increases more than proportionately as concentration increases, so the estimated SCC tends to increase over time (Agrawala et al, 2010;

CGSP, 2013). However, the level of CO_2 concentration is affected by mitigation efforts to slow down the rate at which CO_2 accumulates.

A key element of IAMs is how individuals and society maximise consumption utility over time, considering society's rate of time preference (or the social discount rate). The time profile of the discount rate affects the level of mitigation (and adaptation if relevant and available) adopted in the IAMs. The topic of discount rates will be discussed in detail in Chapter 4 but Table 3.1 shows that the discount rate used in a number of IAMs varies between 1.4% and 6%.

Table 3.1. **Discount rate used in various Integrated Assessment Models**

	Model	Discount rate (%)
	Welfare maximisation models	
DICE	**D**ynamic **I**ntegrated **C**limate-**E**conomy Model (developed by William Nordhaus)	5.5
MERGE	**M**odel for **E**stimating the **R**egional and **G**lobal **E**ffects of Greenhouse gas reduction (developed by Stanford University)	5.0
FUND	Climate **F**ramework for **U**ncertainty, **N**egotiation, and **D**istribution (developed by Richard Tol)	5.0
WITCH	World Induced Technical Change Hybrid model	n/a
	Computable General equilibrium model	
EPPA 4.0	**E**missions **P**rediction and **P**olicy **A**nalysis Model Version 4.0 (developed by MIT)	4.0
Worldscan	Developed by CPB Netherlands Bureau for Economic Policy Analysis	3.0 to 6.0
GTEM	Global Trade and Environmental Model (developed by Australia Department of Agriculture, Fisheries and Forestry)	5.0
	Other	
PAGE 2002 (Hope)	**P**olicy **A**nalysis of the **G**reenhouse **E**ffect used in Hope (2006 & 2008)	3.0 – 4.7
PAGE 2002 (Stern)	**P**olicy **A**nalysis of the **G**reenhouse **E**ffect used in Stern (2006)	1.4

Note: The discount rate used in different models might have changed since the review carried out by Weisbach and Moyer.

Source: Weisbach and Moyer (2010).

Figure 3.4 illustrates the impact of different discount schedules[2] has on the estimated climate change damages in present values using the Dynamic Integrated Climate-Economy model developed by William Nordhaus (Phan, 2011).

Under each set of discount rate assumptions, the level of mitigation implementation could be different (as indicated by the difference between the 'no control' estimate and the 'optimal policy' estimate[3]) and therefore the final CO_2 concentration would also be different. Although it is difficult to infer the corresponding carbon value from Figure 3.4 a declining discount schedule generally yields higher future carbon values in present value terms than that of a constant (but high) discount schedule (Chapter 4). As discussed in the literature, the assumptions such as the discount rate used in many IAMs are different, therefore results between different models can vary substantially (Kane, 2012).

Figure 3.4. **Examples of present value of climate change damage using declining discount rate schedule**

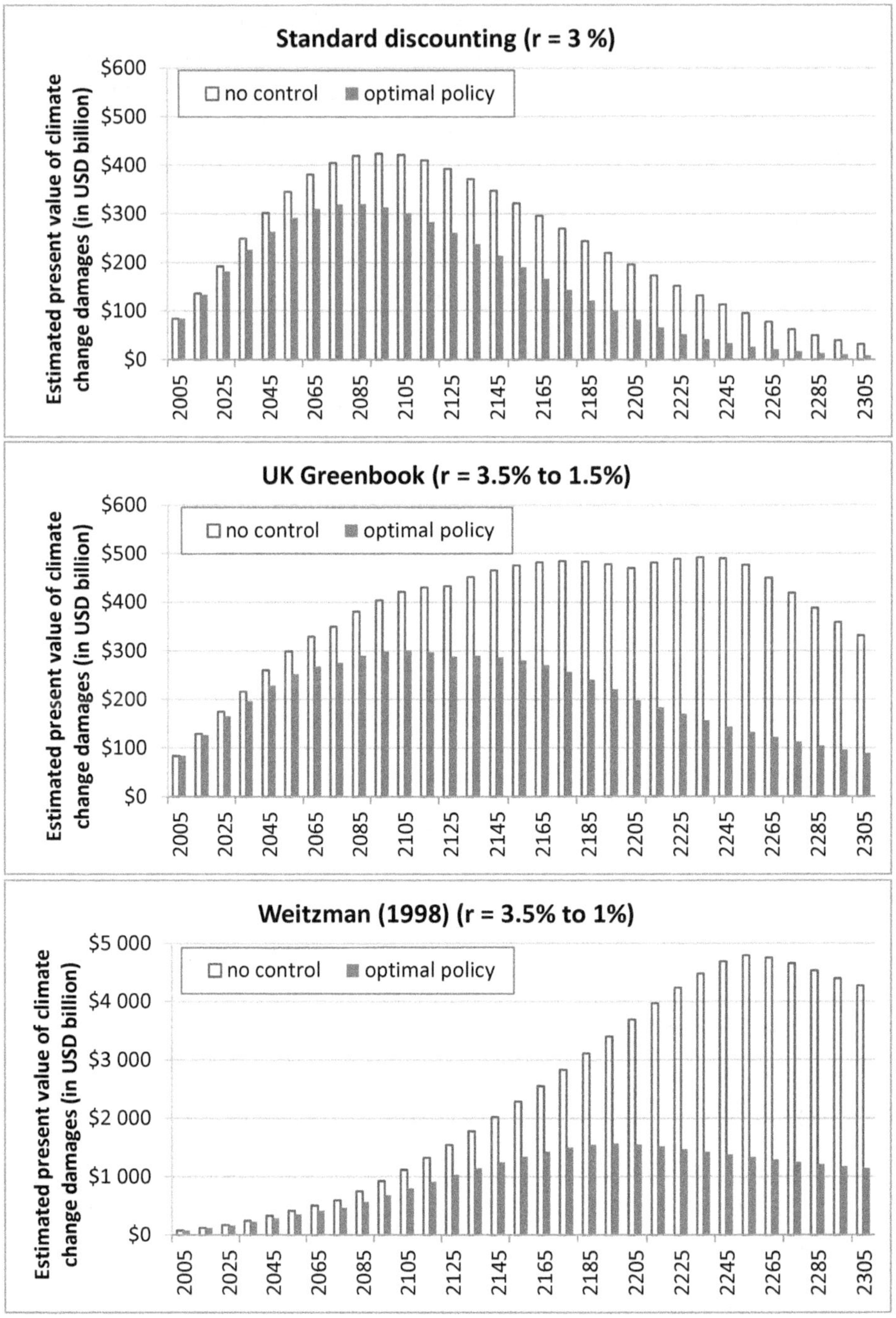

Standard discounting (r = 3 %)

UK Greenbook (r = 3.5% to 1.5%)

Weitzman (1998) (r = 3.5% to 1%)

Source: Phan (2011).

International comparison

This section briefly summarises carbon valuation practices from the partial survey of OECD member countries. Box 3.1 provides more details on each country's approach.

SCC-based estimates of carbon value vary between countries using the same approach as there is no agreement on what inputs or scenarios should be used in IAMs. As noted earlier, Tol (2008) (cited in Mandell, 2013) reviewed 211 SCC estimates from 47 studies using different IAMs and found a wide distribution of estimates, from EUR -1 to EUR 451 per tonne of CO_2.

Carbon values derived from the abatement cost approach also differ across countries and sectors because the cost of reducing CO_2 emissions depends on the social, economic and technological circumstances in each country and sector, the scope and opportunities for emission reduction, as well as on the policy instruments used to trigger the emission reductions (Ireland, 2004 and OECD, 2013a).

Differences between countries exist, not just in how they estimate the value of carbon, but also in how they value carbon over time. In some countries, the carbon value increases over time, in others it does not (Table 3.2). The picture looks somewhat different when the carbon values are discounted at each jurisdiction's chosen discount rate schedule (Figure 3.5). Even with a discount rate that would be considered moderate in most investment contexts (e.g. 3-5% per year), the discounted future carbon value will be substantially lower than the undiscounted value. The impact of discounting is more pronounced when the carbon value is low and the discount rate is high (e.g. in NZ's USD 28 per tonne per CO_2 in 2010 will fall to USD 2 by 2050 using a 8% discount rate).

Table 3.2. **Carbon value used in transport project appraisal in different countries**
(expressed in USD 2013 value per tonne of CO_2)

USD/tCO_2	Approach adopted	2010	2020	2030	2040	2050
France	Abatement cost	39	123	295	Grows at 4.5% p.a.	
U.K. (non-traded)	Abatement cost	40-122	47-142	55-164	106-319	158-473
Norway	Abatement cost	**Before 2015:** 24	**2015-2030:** interpolated	**After 2030:** 91		
The Netherlands	Abatement cost	12.6 to 194.90				
Germany	Damage and abatement costs	**2010:** 54-161		**2030:** 94-288		**2050:** 174-523
U.S.	Damage cost	12-97	13-140	18-174	23-209	28-241
New Zealand	Damage cost	34.4				
Japan	Damage cost	25.7				
Sweden	Fuel tax on CO_2	**For investments < 10 years:** 128			**For investments >10 years:** 172	

Notes: All carbon values were first inflated to 2013 prices in domestic currency using GDP deflators and then converted to USD using PPP conversion factors. GDP deflators and PPP data are sourced from OECD Statistics. A conversion factor of 3.67 is used to convert tC to tCO_2 (for Japan).

Source: OECD survey of monetary carbon values in selected member countries.

In addition to the discount rate, other CBA practices such as evaluation horizon and the estimation of the residual project value can have a major influence on the final value of CO_2 used in CBA. Further work in this area is being conducted as part the OECD's international survey of monetary value of carbon.

According to this survey, only a few of the countries have established common guidelines across all sectors and types of projects for compulsory CBA. For countries that have not established a standardised practice, different carbon values are sometimes used by different sectors within the same jurisdiction. In New Zealand, for instance, the transport sector carbon value is set at NZD40/tCO_2e (2004 prices) but their health and environmental agencies sometimes use carbon market prices or other measures. In Sweden, there were also reports of different values used in different sectors due to a lack of guidance provided. This practice is problematic because it can lead to inconsistent climate policy being adopted by different agencies.

One of the reasons why different sectors have different carbon values is if the abatement cost approach is applied. In this case, differences in each sector's capacity for further CO_2 reduction and the availability of new emission reduction technologies within that sector will affect the relevant carbon value. However, the interagency variations appear to be largely due to a lack of communication between ministries and agencies, rather than intentional disjunction.

Since the policy actions in one sector can often affect the outcomes of other sectors, the use of inconsistent carbon values across sectors can hinder correct assessment of the overall national impacts. For the purpose of assessing the benefits of alternative approaches to mitigation, it would be preferable for different sectors to use the same carbon value within the same jurisdiction and for countries to use relatively similar carbon values. This would ensure consistent assessment of the national impacts and enable more efficient policy mix to be adopted by the jurisdiction.

Figure 3.5. **Effects of discounting on carbon value by country**
(expressed in USD 2013 value per tonne of CO2)

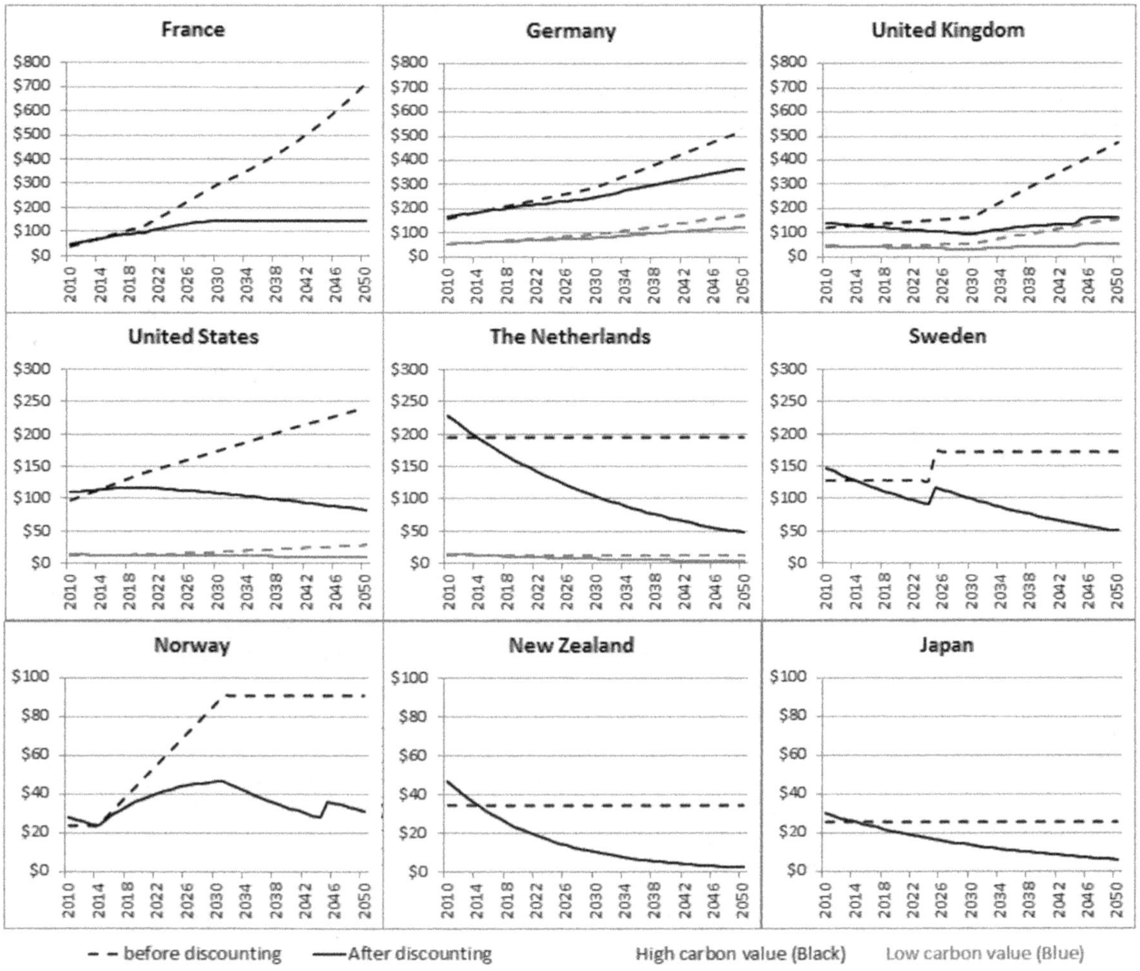

France | Germany | United Kingdom | United States | The Netherlands | Sweden | Norway | New Zealand | Japan

- - before discounting —After discounting High carbon value (Black) Low carbon value (Blue)

Source: ITF calculations based on OECD survey of selected member countries (Tables 2 and 7).

Box 3.1. **Carbon value: International practices**

Japan

In Japan, a working group of Government Committee on Project Evaluation Method was set up in 2008 to review their project evaluation method. One of the areas they looked into was the valuation of environmental impacts. After considering the common approaches for establishing carbon value, the working group recommended the use of the damage cost approach. Based on the results of Tol (1999), the working group recommended that one tonne of carbon be priced at JPY 10,600,[4] with a range of 50% (JPY 5,300/tC) and maximum 200% (JPY 21,200/tC). More details of Japan's approach are provided in Annex A.

Netherlands

Because climate change effects and the associated social cost of carbon are highly uncertain and difficult to estimate using the damage cost approach, the Netherlands uses the abatement cost approach to value CO_2 emission. For 2010, the recommended carbon value[5] ranged from EUR 10 and EUR 155 with a median value of EUR 78 per tonne of CO_2 (Schroten et al., 2014). The upper limit of EUR 155 is based on Kuik et al. (2008) and is associated with a very strict target of 450 parts per million. The lower limit is associated with the current EU policy target of 20% reduction of greenhouse gases in 2020 compared to 1990. All carbon values are expressed in 2010 price level. More details of the Netherlands' approach are provided in Annex B.

Sweden

Prior to 2012, Sweden used the abatement cost approach to determine carbon value. Since then, for short-term projects (with evaluation period of less than 10 years), the carbon value[6] is set at 1.08 SEK per kg of CO_2e emitted in 2010 prices, based on the fuel tax[7] on CO_2. For long-term projects (with evaluation period of 40 years or longer), the carbon value is set at 1.45 SEK per kg of CO_2e emitted in 2010 prices.

Germany

Germany established their carbon value based on the damage cost approach. The current carbon value is set at EUR 80 per tonne of CO_2 in 2010 prices. For longer term effects, it is set at EUR 145 for emissions in 2030 and EUR 260 for emissions in 2050. UBA (2012) also provides lower and upper values for sensitivity testing (Table 3.3). More details of Germany's approach are provided in Annex C.

Table 3.3. **UBA recommendation for carbon value (EUR 2010 / t CO_2)**

	short term 2010	medium term 2030	long-term 2050
minimum value	40	70	130
central value	80	145	260
maximum value	120	215	390

Source: German Federal Environment Agency (UBA, 2012).

New Zealand

As part of a Land Transport Pricing Study conducted in 1996, the New Zealand (NZ) Ministry of Transport looked at the average damage cost of carbon dioxide emissions and established a social cost value of NZD 30 per tonne of CO_2. This value was later adopted by the NZ Transport Agency and updated to NZD 40 in 2004 dollars. The estimate was also converted to cents per litre of fuel used and percentage of vehicle operating costs. Since then, one of these three measures has been used in transport appraisals. In non-transport sector, carbon prices or other measures are sometimes used.

Box 3.1. **Carbon value: International practices** *(cont.)*

United States

In 2008, the United States (US) Federal Government established an interagency working group to develop a SCC value to be used in CBAs conducted by all the departments and agencies in the US Government.[8] The SCC estimates are based on three commonly used IAMs:

- the Policy Analysis of the Greenhouse Effect, developed in 1991
- the Dynamic Integrated Climate and Economy, developed by William Nordhaus in 1990; and
- the Climate Framework for Uncertainty, Negotiation, and Distribution, developed by Richard Tol in the early 1990s.

These models calculate the damage associated with global warming by using CO_2 emissions projections as an input. To account for the unknown long-term effects, the models use five different socioeconomic scenarios, based on the results of a Stanford Energy Modelling Forum exercise, and three discount rates – 2.5%, 3% and 5% (chosen by the interagency group to be applied in operating the IAMs). Based on these variables, each model produces 15 separate SCC distributions for a given year (45 estimates in total).

The distributions from each model and scenario were equally weighted and combined to produce three separate probability distributions for SCC in a given emissions year, one for each of the three discount rates. From the three distributions, the interagency group selected four final values in order to produce a range that reflects sensitivity to discount rate assumptions and uncertainty in climate impacts: the average SCC at each discount rate (2.5%, 3%, and 5%), and the 95th percentile at a 3% discount rate, representing higher than expected economic impacts further out in the tails of the distribution. The SCC estimates were originally established in 2010, and were updated in 2013. They are currently applied in CBAs of all major U.S. regulatory policies (USG 2013 and Greenstone et al., 2013).

Table 4.6. **The US social cost of carbon (2007 USD/tCO_2)**

Discount Rate	5.0%	3.0%	2.5%	3.0%
Year	Avg	Avg	Avg	95th
2010	11	32	51	89
2015	11	37	57	109
2020	12	43	64	128
2025	14	47	69	143
2030	16	52	75	159
2035	19	56	80	175
2040	21	61	86	191
2045	24	66	92	206
2050	26	71	97	220

Source: US Government (2013).

One example of the SCC being applied in the United States was a Federal Government CBA of the emissions and fuel efficiency standards for light-duty vehicles. The analysis indicated that the regulation had a benefit of more than USD 170 billion (2007 dollars) from carbon reduction alone (Greenstone et al., 2013).

France

France uses the abatement cost approach to estimate the carbon value in CBAs. To calibrate the marginal abatement cost curve, simulations were conducted based on three different models:

- POLES, developed in the early 1990s at the Institute of Energy Policy and Economics IEPE (now LEPII-CNRS) by Criqui (1996);

Box 3.1. **Carbon value: International practices** *(cont.)*

- GEMINI E3, developed in 1994 at the Energy Atomic Agency under the supervision of Alain Bernard; and
- IMACLIM-R, developed by the Centre International de Recherche sur l'Environnement et le Développement.

The simulations looked at three different scenarios on the constraints of CO_2 emissions, based on different assumptions of international commitments. Nine different carbon values were derived as a result. By synthesising these multiple elements of information and knowledge, the final chosen carbon value in France was EUR 32 per CO_2 ton in 2010, increasing (at 5.8% per year) to EUR 100 in 2030[9]. After 2030, the increase in carbon value is assumed to follow the Hotelling rule[10] (i.e. same growth rate as discount rate or 4.5%) and reach EUR 240 by 2050. All estimates are expressed in 2010 monetary value.

United Kingdom

Prior to 2009, the UK Government used what they called the shadow price of carbon, an estimate that is based on the SCC for a given stabilisation goal, with adjustments for marginal abatement cost and the willingness-to-pay for reductions in carbon emissions (Defra, 2007).

Following a review in 2009, the UK Government adopted a "target-consistent" approach to carbon valuation. For sectors that are covered by EU-ETS (EU Emissions Trading System), the market price of CO_2 has been used. This has been considered as the market value of the marginal abatement cost, which proxies an optimal unit price of CO_2 reduction. For sectors that are not covered by EU-ETS, the marginal abatement costs are developed separately, based on their own cost estimation model. While the initial values for the two markets differ, they are expected to converge in the long run under a carbon trading system (DECC, 2009).

In 2012, DECC's methodology for producing short-term traded carbon values was updated. Since then, short-term traded carbon values are based on expected future market prices. The values were updated in 2013 (Table 3.7).

Table 3.7. **The UK carbon value (2013 GBP Per tonne of CO_2e)**

	Traded			Non traded		
2013 £	**Low**	**Central**	**High**	**Low**	**Central**	**High**
2010	12	12	12	28	57	85
2015	0	4	18	30	61	91
2020	0	5	26	33	66	99
2025	19	41	70	36	71	107
2030	38	76	114	38	76	114
2035	56	112	168	56	112	168
2040	74	148	222	74	148	222
2045	92	183	275	92	183	275
2050	110	219	329	110	219	329

Source: UK DECC (2013a).

Norway

Based on the Climate Cure 2020 modelling results of achieving Norwegian climate goals by 2020, the Norwegian Public Roads Administration recommended the carbon value for the years to 2015 to be NOK 210 per tonne of CO_2, and NOK 800 of for the years from 2030[11]. The values between 2015 and 2030 are interpolated based on NOK 210 and NOK 800[12].

Source: Smith, S. and N. A. Braathen (2015).

Notes

1. Negative values are due to an initial positive impact on crops, etc. (Mandell, 2013).

2. The discount rates used were (i) a constant 3%; (ii) the declining discount rate based on UK Treasury Green Book's discounting schedule – 3.5% for the first 30 years, 3% for next 40 years, 2.5% for next 50 years, 2% for next 75 years and 1.5% for next 100 years; and (iii) the declining discount rate schedule as suggested by Weitzman (1998) – 3.5% for the first 20 years, 2% for the next 50 years and 1% for the next 230 years.

3. While this example used in Phan (2011) focuses on policy optimization, other IAMs focus on welfare maximisation and those that look at general or partial equilibrium effects (Ortiz and Markandya, 2009).

4. This roughly equals to USD 25.7 (2013 value) using GDP deflators and PPP conversion.

5. These can be expressed in USD 12.6 (low), USD 98.1 (median) and USD 194.9 (high), using GDP deflators and PPP conversion to convert to 2013 USD (Schroten et al., 2014).

6. For evaluation period of less than 10 years, the carbon value is equivalent to USD 128 per tonne. For evaluation period of 40 years or longer, the carbon value is around USD 172 per tonne. Both figures are expressed in 2013 value using GDP deflators and PPP conversion.

7. In Sweden, the fuel tax contains two components – an energy tax and a CO_2 tax. The energy tax is levied on most fuels based on their energy contents. The aim is primarily fiscal but also to improve energy efficiency and increase the use of renewable energy. Tax rates are higher for motor fuels in order to take account of external effects such as noise, congestion and road wear from traffic. The CO_2 tax was introduced in Sweden in 1991. Over the years the tax rate has been significantly increased, in order to take account of the need to fight climate change. At present, the general CO_2 tax rate corresponds to more than USD 125/tonne.

8. See Kopits et al. (2013).

9. Sources: Quinet (2013) and CGSP (2013).

10. Hotelling rule states that the use of an extractive resource should be distributed over time optimally at all points in time. Therefore, the "present value of the resource price must rise over time at the discount rate" (Heal, 1997).

11. These are equivalent to USD 24 and USD 91 (2013 value) using GDP deflators and PPP conversion.

12. Norwegian Ministry of Finance (2012), p.145.

References

Agrawala, S. et al. (2010), "Plan or react? Analysis of adaptation costs and benefits using integrated assessment models", *OECD Environment Working Papers*, No. 23, OECD Publishing, Paris, http://dx.doi.org/10.1787/5km975m3d5hb-en.

Anhoff, D. and Tol, R.S.J. (2010), "On international equity weights and national decision making on climate change", *Journal of Environmental Economics and Management*, Vol. 60, 14-20.

Climate Cure 2020 (2010), "Measures and instruments for achieving Norwegian climate goals by 2020", Report number TA2678/2010, Published by Climate and Pollution Agency, Oslo, June, www.miljodirektoratet.no/old/klif/publikasjoner/2678/ta2678.pdf.

Commissariat général á la stratégie et á la prospective (CGSP) (2013), "Évaluation socioéconomique des investissements publics : Rapport de la mission présidée par Émile Quinet", Sept.

Criqui, P., et al (1996), "POLES 2.2.- Bruxelles : Communautés Européennes", DG XII, Programme Joule II, Dec.

DCCEE (2011), "Estimating the cost of abatement: Framework and practical guidance", October, Department of Climate Change and Energy Efficiency, Australian Government.

DECC (2013a), "Green Book supplementary guidance: Valuation of energy use and greenhouse gas emissions for appraisal", September, Department of Energy and Climate Change, United Kingdom.

DECC (2009), "Carbon valuation in UK policy appraisal: A revised approach", July, Department of Energy and Climate Change, United Kingdom.

Defra (2007), "The social cost of carbon and the shadow price of carbon: what they are, and how to use them in economic appraisal in the UK", December, Department for Environment, Food and Rural Affairs, United Kingdom.

Gouvernement du Québec (2014), "Auction of Québec greenhouse gas emission units on August 26, 2014: Summary report results", Québec.

Greenstone, M., E. Kopits, and A. Wolverton (2013), "Developing a social cost of carbon for US regulatory analysis: A Methodology and Interpretation", *Review of Environmental Economics and Policy*, Vol. 7(1), pp. 23-46.

Heal, G (1997), "Discounting and climate change: An editorial comment", *Climate Change*, Vol. 37(2), pp.335-343.

Hope, C. (2006), "The marginal impact of CO_2 from PAGE2002: an integrated assessment model incorporating the IPCC's five reasons for concern", *Integrated Assessment*, Vol. 6(1), pp. 19-56.

Hope, C. (2008), "Discount rates, equity weights and the social cost of carbon", *Energy Economics*, Vol. 30(3), pp. 1011-1019.

Ierland, W. (2004), "A user's guide for economic instruments in domestic and international climate change policy: What role can they play in a Belgian climate change strategy?" Federal Planning Bureau, Belgium.

IPCC (2007),"Climate Change 2007: Mitigation. Contribution of working group III to the Fourth assessment report, [B. Metz et al. (eds.)]", Cambridge University Press, Cambridge, United Kingdom and New York, NY, USA.

Kane, M. (2012), "Disagreement and design: An arbitration of the climate change and intergenerational discounting debate", New York University Law and Economics Working Papers, Paper 306 http://lsr.nellco.org/nyu_lewp/306.

Kopits, E., Marten, A. L. and Wolverton A. (2013), "Moving forward with incorporating catastrophic climate change into policy analysis", National Centre for Environmental Economics, US Environmental Protection Agency.

Kuik, O. et al. (2008), "Methodological aspects of recent climate change damage cost studies", *The Integrated Assessment Journal*, Vol. 8(1), pp.19-40.

Mandell, S. (2013), "Carbon emissions and cost-benefit analysis", *International Transport Forum Discussion Papers,* No. 2013/32, OECD Publishing, Paris. DOI: http://dx.doi.org/10.1787/5jz40rmnqzvc-en

Norwegian Ministry of Finance (2012), "Cost-benefit analysis", Official Norwegian Reports NOU (2012), October, Norway.

OECD (2013a), *Effective carbon prices*, OECD Publishing, http://dx.doi.org/10.1787/9789264196964-en.

OECD (2012), *Mortality risk valuation in environment, health and transport policies*, OECD Publishing, Paris, http://dx.doi.org/10.1787/9789264130807-en.

Ortiz, R. A. and Markandya, A. (2009), "Integrated impact assessment models of climate change with an emphasis on damage functions: A literature review", *Basque Centre for Climate Change working papers*, 2009-06.

Parry, I., de Mooij, R. and Keen, M. (2012), "Fiscal policy to mitigate climate change – guide for policymakers", International Monetary Fund, Washington DC.

Phan, D (2011), "The use of time declining discount rates in climate change projects", MSc Thesis, Environmental Economics and Natural Resources, January.

Pindyck, R.S. (2013), "The climate policy dilemma", *Review of Environmental Economics and Policy, Association of Environmental and Resource Economists*, Vol. 7(2), pp. 219-237, July.

Quinet E. (2013), "Factoring Sustainable Development into Project Appraisal: A French View", *International Transport Forum Discussion Papers*, No. 2013/31, OECD Publishing, Paris. DOI: http://dx.doi.org/10.1787/5jz40rncm222-en.

Schroten, A. et al. (2014), "*Externe en infrastructuur - kosten van verkeer: Een overzicht voor Nederland in 2010*", Delft, CE Delft.

Smith, S. and N. A. Braathen (2015), "Monetary Carbon Values in Policy Appraisal: An Overview of Current Practice and Key Issues", *OECD Environment Working Papers, No. 92*, OECD Publishing, Paris. DOI: http://dx.doi.org/10.1787/5jrs8st3ngvh-en

Stern, N. (2006). "Stern Review on The Economics of Climate Change. Executive Summary". HM Treasury, London.

Tol, R.S.J. (2008), "The social cost of carbon: Trends, outliers and catastrophes." *Economics, The Open-Access, Open-Assessment E-Journal*, Vol. 2, 1-24.

Tol, R.S.J. (1999), "The marginal costs of greenhouse gas emissions", *The Energy Journal*, Vol. 20(1), pp.61-81.

UBA (2012), "Best practice cost rates for air pollutants, transport, power generation and heat generation – Annex B to Economic valuation of environmental damage – methodological convention 2.0 for estimates of environmental costs": Dessau-Roßlau.

USG (2013), "Technical update of the social cost of carbon for regulatory impact analysis under executive order 12866", Interagency Working Group on Social Cost of Carbon, United States Government, November.

Weisbach, D. and Moyer, E. (2010), "Discounting in integrated assessment", http://www.rff.org/Documents/Discounting%20in%20Integrated%20Assessment.pdf.

Weitzman, M. L. (1998), "Why the far-distant future should be discounted at its lowest possible rate", *Journal of Environmental Economics and Management*, Vol. 36(3), pp.201-208.

World Bank (2014), "State and trends of carbon pricing, 2014", Washington DC, World Bank.

Chapter 4

Uncertainty and transport appraisal of climate change effects

This chapter will first clarify the difference between risk and uncertainty. It will then discuss climate change-related uncertainties and possible approaches to deal with them in transport appraisal processes.

Risk and uncertainty

Climate change effects are cumulative. Without mitigation, the expected damages in the future could be catastrophic and irreversible. At the same time, the potential to adapt to climate change is poorly understood. For both reasons, estimation of the environmental and economic effects of climate change (with and without a policy) in a distant future is subject to tremendous uncertainties.

To ensure long-term effects are appropriately captured in transport decision-making, it is necessary to consider how risk and uncertainty in the context of climate change impacts can be treated in transport appraisal. Risk and uncertainty are often used interchangeably to refer to unknown events. However, there are subtle differences between these terms.

- Unknown refers to the state where the occurrence of a given event cannot be known prior to the event. This can refer to both risk and uncertainty.
- Risk refers to unknown outcomes with well-defined probabilities and typically with reference to the probability of incurring a loss. Risk can be loosely referred as the *known unknowns*. A stylised example is the probability (or statistical average) of getting a 3 when rolling a fair dice is 1/6, but the number that will turn up on any particular roll of the dice is unknown.
- Uncertainty refers to unknown outcomes with unknown probabilities. Therefore, uncertainty can be loosely referred as the *unknown unknowns*. In this case, the probability of occurrence cannot be determined (Knight, 1921). Long-term climate change effects on the environment and economy fall largely in this category. Scientific evidence suggests the effect of climate change could be catastrophic, and that the timing of future climate change, the probability of a catastrophic event and the extent of its effects are all unknown.

Transport infrastructure appraisals are subject to both risk and uncertainty. With estimates of the probability of occurrence and likely outcomes, preventative measures can be implemented to mitigate certain risks. For example, to address the risk of damage to a man-made structure (e.g. coastal highways or railway tracks) due to sea wall erosion, the existing sea wall could be strengthened with improved materials or features.

Quantifiable risk can also be incorporated in assessment practice by using a probability distribution of likely values rather than using a simple average or certain percentile point. In the case of risk neutrality, the equivalent value of the distribution is simply the average value of the distribution. However, in the presence of risk aversion there will be a risk premium in addition to the expected value. When calculating the net present value over a time period, the risk premium associated with risk aversion may be introduced as an additive correction to annual expected value, and discounted at a risk free rate. Another way to deal with risk is to adjust the discount rate used in assessment, as suggested by Quinet (2013). This will be discussed further in Chapter 4.

In infrastructure project appraisal, uncertainty exists both in the short and long-term, and on both the cost and benefit sides. Some of the transport investment uncertainty relates to costs, such as construction cost and operational costs. What has been more problematic for transport project appraisal is the uncertainty around long-term traffic demand. Demand projections[1] typically rely on many uncertain factors (such as population and economic growth; technological and land-use developments; lifestyle changes and future fuel prices) which are difficult to predict. However, transport demand uncertainties are considerably more tractable because scenarios can be constructed to test the robustness of demand projections and the resulting impacts can be calculated accordingly. It is much harder to deal with uncertainties related to climate science and its economic impacts because it is very difficult to assign a robust probability distribution to such events. This subject is taken up in the next chapter.

Climate change-related uncertainties

Catastrophic events related to climate change and the potential socio-economic impacts of extreme climate change are extremely difficult, if not impossible, to quantify. Scientific knowledge is currently insufficient to predict the magnitude or probability of such impacts. Economic theory has little to say about the potential impact of large temperature increases, and the potential for adaptation compounds uncertainty over impacts. Figure 4.1 provide a stylised representation of the key uncertainties around estimating the climate change impacts. This shows a continuous sequence of stages. Mitigation measures can affect the emission outcomes (level and concentration) and the flow-on impacts on the climate, while adaptation measures affect the outcomes for the natural and man-made systems and so forth.

Figure 4.1. **Uncertain impacts of climate change**

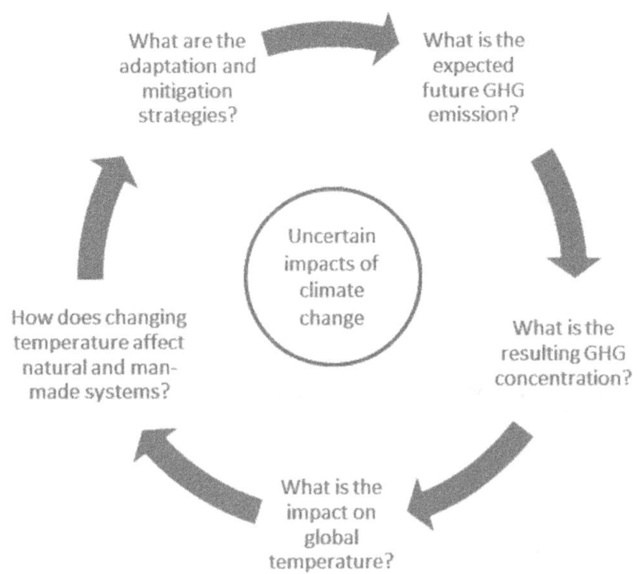

These uncertainties can be categorised into 'scientific uncertainty' and 'socio-economic uncertainty' (Heal and Millner, 2014). Scientific uncertainty reflects our incomplete knowledge of the climate system. Despite efforts that have been made to develop complex climate models based on scientific evidence and observations, such work is still imperfect. As shown in Figure 4.2, the probability distributions of modelled average surface warming, resulting from a doubling of the atmospheric concentration of CO_2, vary significantly between models. There has been disagreement among economists and scientists about the nature and the extent of the uncertainties, the measurement of social welfare and key behavioural or policy parameters that affect it (Pindyck, 2013).

Socio-economic uncertainty, on the other hand, relates to what is unknown about how societies and economies will function in the future and how they respond to climate change. As temperature rises, societies will be subject to changing weather patterns. However, the extent of the damage and cost to society from these changes will be highly dependent on future infrastructure, technology, how economies and societies operate, and other future policies. Society's ability to adapt to climate change will also have an impact on the value associated with mitigation policies. The more successful is adaptation to climate change the lower the value of policies that aim to cut emissions. It is clear that we have the capability to adjust our life and economic activity to some extent in response to climate change. The unknowns are just how much adaptation is possible in the timeframe considered, and what climate we will be adapting to.

Figure 4.2. **Estimates of the probability distribution for climate sensitivity**

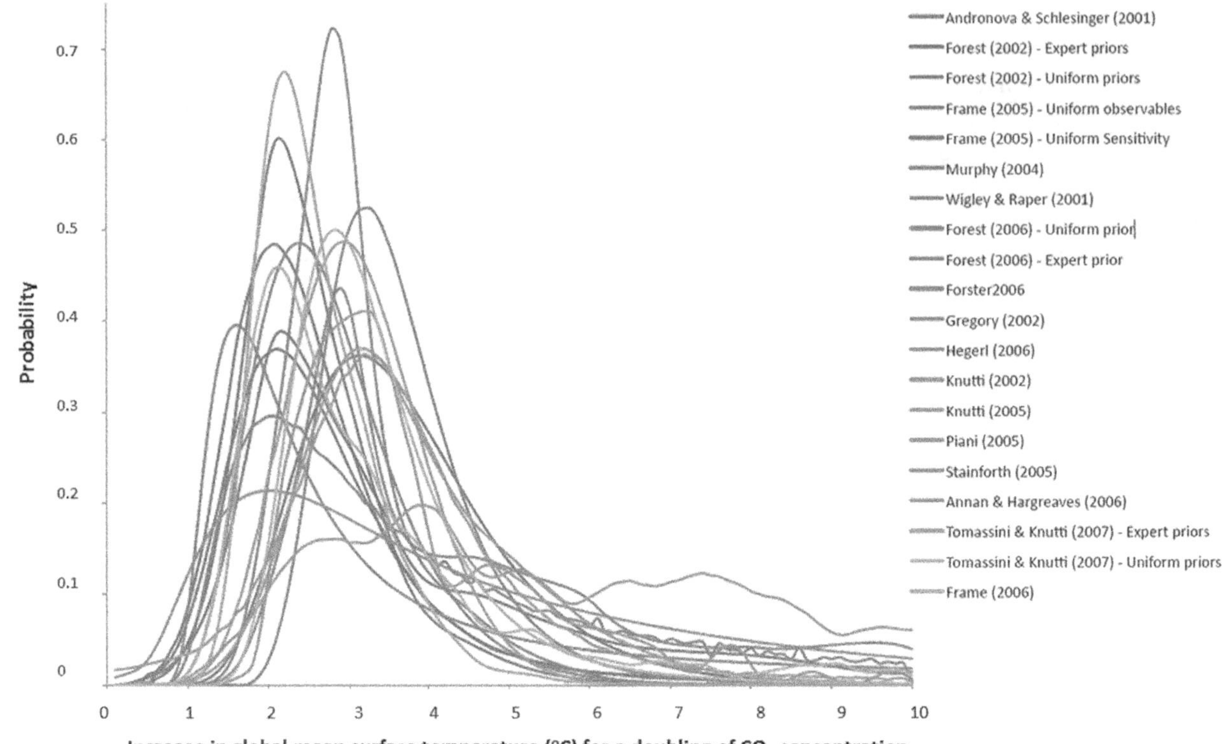

Source: Millner et al. (2012).

Dealing with uncertain long-term impacts in transport appraisal

A key uncertainty associated with transport policies to mitigate climate change effects relates to the occurrence of catastrophic events. Catastrophic events, in relation to the economic analysis of climate change, are characterised by a low probability of occurrence but a high potential of severe damages. This is where the difficulty arises in handling catastrophic impacts in assessments. There is no clear scientific evidence that catastrophes will occur in the future but neither is there any scientific basis for them not occurring. Likewise, the magnitude of damages and their probabilities are unknown and may be largely unknowable. This characteristic has important implications for policy appraisal.

A number of recent papers debated the theoretical framework for analysing catastrophic impacts (e.g. Weitzman, 2009 and Pindyck, 2011). But the academic community has yet to come to a consensus on how catastrophes should be incorporated into assessment practice, particularly in CBA. Weitzman (2009) introduced a so-called "Dismal Theorem", which asserts that the impact of catastrophic events such as climate change is likely to follow a fat-tailed probability distribution, meaning that its upper tail declines to zero only very slowly. The implication of this assertion is that the probability of catastrophic events will be large enough to make expected marginal utility (which declines as consumption grows) effectively infinite (Pindyck, 2011). This means the expected gain from any climate change mitigation policy would be unbounded. This theory challenged the results of traditional climate change-related CBA that uses thin-tailed distribution (e.g. normal distribution or exponential function) in the IAMs, on the ground that these models will underestimate the gains from abatement.

Pindyck (2011 & 2013) argued that even as consumption approaches to zero (which implies death), the marginal utility of consumption would become very large but should still be finite (e.g. approaching multiples of the value of statistical life). Once an upper bound is added to marginal utility, Pindyck argued that expected marginal utility will be finite irrespective of the shape of the damage function. Pindyck's papers also demonstrated that the value of climate change abatement policy depends on two equally important factors – the probability distribution of the catastrophic events and the impact of a catastrophic outcome.

Pindyck tested different probability distributions and calculated the willingness-to-pay to mitigate climate change effects and found that thin-tailed distribution can actually yield a higher marginal utility and willingness-to-pay than a fat-tailed one. This is because when different distributions are calibrated to the same mean and standard deviation, the thin-tailed distributions actually used in IAMs have more mass at the upper tail than that of the fat-tailed distribution (Pindyck, 2011). In other words, the aggregate probability for a thin-tailed distribution is higher in the upper tail. This means higher expected marginal utility for the thin-tailed distributions. Another related point made by Pindyck is that due to the presence of other potential catastrophes (such as nuclear wars or pandemics of a serious disease) the willingness to pay (WTP) for avoiding serious climate change would have to fall when all other catastrophes are considered simultaneously.

The key implication of these arguments is that if the expected gain from climate change mitigation policy is bounded and there are multiple types of catastrophes, the WTP, i.e. the carbon value, would not be unbounded. Therefore, there is still a role for IAMs and CBA and these tools continue to be relevant for analysing the likely climate change impacts. Pindyck (2013) recommended the use of "plausible" estimates of probability of various large economic impacts from various "plausible" temperature change scenarios in climate change policy research with application of standard CBA techniques.

As a modelling tool to estimate climate change impacts, IAMs play an important role in understanding the true costs of climate change. Inadequate modelling of catastrophic events in IAMs has been a common criticism. Kopits et al. (2013) recommended several near-term modelling improvements such as developing better mapping of how physical changes relate to economic damages and improving representation of the key physical future outcomes through which economic consequences are most likely to be significant and incorporating more recent findings and data from scientific literature into IAMs. The same authors also recommended improved communication between transport practitioners and IAM modellers to ensure any limitations of the modelling results are appreciated in the appraisal and subsequent decision-making process.

Under traditional transport appraisal frameworks, unknown impacts that are likely to occur in the distant future are typically excluded from the analysis. This is based on the view that once such future values are discounted back to present values they will be close to zero. However, in assessing climate change effects this may not be appropriate and adjustments to both discount rates and future carbon values have been used to encompass such potential impacts. Carbon values increase in real terms over time in some assessment methodologies and in some approaches assessed using a very low or declining discount rate (this is discussed in Chapter 4).

Climate change-related policy or investment decisions include adaptation and mitigation aspects. Tools that supplement conventional CBA when considering long-term uncertain impacts for policy decisions can vary between the types of decision at hand.

- ***Impact assessments*** – For benefits or costs that are difficult to monetise, individual impact analysis can be considered to supplement traditional CBA. This approach is common with effects such as biodiversity, distributional impacts and changes in landscape. However, as the number of non-monetised factors increase, the level of influence CBA has will weaken since the result represents smaller part of costs and benefits being considered in policy appraisal.

- ***Cost-effectiveness analysis (CEA)*** – Given the uncertainty involved in calculating the carbon value for a CBA, CEA is often used (such as Ackerman et al., 2009 and Official Norwegian Reports, 2012). A CEA typically involves comparison of the relative costs of different courses of action that aim to achieve the same outcomes or relative outcomes of courses of action that incur the same costs. The key distinction between a CEA and CBA is that a CEA does not require monetisation of the outcomes. While CEA may be a useful alternative to CBA, it would not provide any indication as to the value for money of a project. Another disadvantage is that a CEA does not provide a measure of the long-term sustainability. Furthermore, ranking policy options using CEA is not sufficient because there are other policy measures for which carbon abatement is not the primary goal (Defra, 2007). Other approaches would still be needed to compare climate change-related projects with other projects.
- ***Sensitivity testing and scenario analysis*** – Sensitivity testing and scenario analysis involve having several choices of data inputs for a CBA or setting up different scenarios for modelling a range of possible impacts. These approaches provide decision-makers a range of estimates and scenarios to consider. In some cases, such an analysis would be sufficient to inform policy decisions. In cases where there is a wide range of possibilities, analysts need to exercise careful judgement when selecting the scenarios to investigate. It is also necessary to provide sufficient explanations to decision-makers to ensure the limitations of the analyses are appreciated.
- ***A real options approach[2] to adaptation policies*** – There is a strong possibility that, over the long-term, the frequency, intensity and nature of extreme weather events may accelerate infrastructure degradation, inducing unexpected repair and maintenance costs, or even to catastrophic failures. A first reflex, from a transport decision-making perspective, would be to enlarge the policy questions: rather than just ask "should we build the project or not?" also address "build now or postpone the decisions for a few years?" (Quinet, 2013). From the project design perspective, infrastructure could be maintained to a high standard to lower its vulnerability to out-of-norm events. Incorporating strategic redundancy in networks could ensure overall network performance would not be too severely impacted even if some assets were to fail. Another possibility is to deliberately under-design parts of the network to "safe-fail" to allow rapid restoration of services.

Mitigation of climate change requires large sunk costs in the near term with highly uncertain benefits occurring in the far future. Different mitigation or adaptation options come with different costs and have different value for money implications. As many mitigation investment options are irreversible, the timing of the investment is crucial. Researchers (such as Hendricks, 1992 and Pindyck, 2000 cited in Kopits et al., 2013) have used a real options framework to examine the characteristics of optimal climate change policy to enable policy makers to learn about the expected damages over time. Based on the assumption that climate change-related uncertainty can be partially resolved over time, many of the real options studies suggested that delaying abatement actions may be optimal[3] (for a review, see Kopits et al., 2013).

Dobes (2008), on the other hand, looked at the strategy to invest in climate change adaptation measures and demonstrated the benefit of using an adaptive investment strategy based on the real options framework. Dobes demonstrated that the real options approach is not just about having strategies to delay decisions until uncertainty is resolved. He used simple examples to demonstrate strategies that could be used to adapt to climate change, including staged investment and more flexible design of infrastructure to allow its scope or coverage to be extended if necessary. An example is the decision to address the potential problem of flooding in low-lying areas: instead of building a full size sea wall, the option would be to engineer a suitable base first to allow an expansion of the sea wall in the future if required. Other real options strategies include the option to abandon or temporarily shut down staged investment, the option to switch the way a demand is met, and the option to adopt a non-asset approach (such as research and development).

Each of these options – deferral, expansion, contraction, abandonment and modification – adds value to a project by allowing decision makers to exploit upside opportunities (e.g. through project expansion) while limiting downside losses (e.g. by abandoning or downsizing projects).

The real options framework is not new[4] and it has been used for investment decisions such as oil field development (Lund, 1999), electricity capacity expansion (Gahungu and Smeers, 2012) and ship investment (Pires et al, 2012). It has not received the same level of attention in the public sector. However, in 2009, real options approach was included in HM Treasury's Supplementary Green Book Guidance[5] for assessing activity that has uncertainty, flexibility and learning potential. More recently, some Australian experts and New Zealand transport officials[6] have been actively investigating the use of the real options approach (using decision tree analysis)[7] in policy and investment decisions.

The real options framework supplements conventional CBA for assisting investment decisions in the face of uncertainty and improves the option development and selection process. Under this framework, CBA is still a useful tool to inform the value for money of projects because all costs and benefits are monetised and compared in a consistent manner.

Uncertainty as described by Knight (1921) refers to circumstances where statistical quantification of the unknowns is not possible. Since the above approaches require assumptions on the probability of occurrence, they can only address risk that can be reasonably reflected by a formal probability distribution.

To ensure the quality of policy and investment decisions, decision-makers need to be informed about how uncertainties (such as those on future demand and economic conditions) affect the estimated costs and benefits of an intervention or investment. This could be done by carrying out an uncertainty assessment, which should be similar to sensitivity testing or scenario analysis but with a focus on assessing the relative project or investment outcomes under different uncertain future states. The assessment will provide decision-makers with explicit information about the uncertainties involved and how they impact on the overall cost and benefit positions. While no robust method to incorporate uncertainty in CBA is currently available, having a separate uncertainty assessment would be of value to decision-makers.

In practice, this would mean providing a likely range of results after considering the probability of occurrence (i.e. consideration of risk) and another wider range of results that consider the impacts of uncertainty. The latter will need to be supported by descriptions of the sources of uncertainty, its determinants and potential impacts.

Notes

1. An example of demand risk is the UK Channel Tunnel Rail Link (which became fully operational in 2007). During the PPP bidding phase in the mid-1990s, consultants consistently over-estimated passenger forecasts. The final passenger forecasts were out by a factor of 3 or more compared to the realised patronage levels. Aside from the possibility of the result of a commercial bidding strategy, the project was the first of its kind and there was no precedent of similar projects anywhere in the world to allow consultants to base their forecasts on (OECD, 2013b).

2. In the financial markets, a financial option gives the bearer the right, but not the obligation, to buy (a put option) or sell (a call option) a financial security in the future, under pre-determined terms and conditions. The real options approach applies this financial options theory to real investments such as roads. The real options approach allows decision makers to retain the right, but not the obligation, to exploit various 'go' and 'no go' opportunities in the future.

3. This is because the expected payoff (the probability weighted average of possible payoffs) from delaying decisions is higher than if actions are implemented before resolving the uncertainty.

4. The real options approach was first mentioned in the literature by Stewart Myers (Myers, 1977) and more recently by Dixit and Pindyck, 1994; Trigeorgis, 1996 and Brennan and Trigeorgis, 2000, etc. (cited in de Neufville, 2003).

5. HM Treasury and Defra (2009), Chapter 3.1.

6. This is based on private communications with ACIL Allen Consulting (Australia) and NZ Ministry of Transport.

7. The decision tree approach traces the evolution of the option's key underlying variables in discrete-time (e.g. Copeland and Tufano, 2004). Each node in the tree represents a possible value of the underlying asset at a given point in time. The option value is calculated iteratively, starting from the final node of each branch of the tree, and then working backwards through the tree towards the first node. The decision tree approach is simple to apply but yet flexible for analysing complex decisions. Other common approach includes the Black-Scholes model (see Luehrman, 1998 and ACIL Tasman 2013).

References

ACIL Tasman (2013), "Real options – Major project development assessment and approvals: use of a real options approach", Prepared for the Productivity Commission, Australia.

Ackerman, F. et al. (2009), "Limitations of integrated assessment models of climate change", *Climatic change*, Vol. 95(3-4), 297-315.

Brennan, M. and Trigeorgis, L. (2000), *Project flexibility, agency, and competition: new developments in the theory and application of real options*, Oxford University Press, Oxford, UK and New York, NY.

Copeland, T. and Tufano, P. (2004), "A real-world way to manage real options", *Harvard Business Review*, March.

de Neufville (2003), "Real options: dealing with uncertainty in systems planning and design", *Integrated Assessment*, Vol. 4(1), pp.26-34.

Defra (2007), "The social cost of carbon and the shadow price of carbon: what they are, and how to use them in economic appraisal in the UK", December, Department for Environment, Food and Rural Affairs, United Kingdom.

Dixit, A. and Pindyck, R. (1994), *Investment under Uncertainty*, Princeton University Press, Princeton, NJ.

Dobes, L (2008), "Getting Real about Adapting to Climate Change: Using 'Real Options' to Address the Uncertainties", *Agenda: A Journal of Policy Analysis and Reform*, Vol. 15(3), pp. 55-69.

Gahungu, J and Smeers, Y (2012), "A real options model for electricity capacity expansion", Robert Schuman Centre for Advanced Studies, Loyola de Palacio Programme on Energy Policy, RSCAS 2012/08.

Heal, G. and A. Millner (2014), "Uncertainty and decision making in climate change economics", *Review of Environmental Economics and Policy*, Vol. 8(1), pp. 120-137.

Hendricks, D (1992), "Optimal Policy Response to an Uncertain Threat: The Case of Global Warming", Unpublished manuscript, Kennedy School of Government, Harvard University.

HM Treasury and Defra (2009), "Accounting for the effects of climate change: Supplementary Green Book guidance", London, United Kingdom.

Knight, F. (1921), *Risk, uncertainty and profit*, Boston, MA: Hart, Schaffner & Marx; Hougton Mifflin Company.

Kopits, E., Marten, A. L. and Wolverton A. (2013), "Moving forward with incorporating catastrophic climate change into policy analysis", National Centre for Environmental Economics, US Environmental Protection Agency.

Luehrman, T (1998), "Investment opportunities as real options: getting started on the numbers", *Harvard Business Review*, July – August.

Lund, M. (1999), "Real options in offshore oil field development", conference paper presented to 3rd Annual Real Options Conference, http://realoptions.org/papers1999/LUND.PDF.

Millner, A., Dietz, S., Heal, G., (2012), "Scientific Ambiguity and Climate Policy", *Environmental and Resource Economics*, Volume 55, Issue 1, pp. 21-46, DOI 10.1007/s10640-012-9612-0.

Myers, S (1977), "Determinants of corporate borrowing", *Journal of Financial Economics*, Vol. 5, pp. 147-175.

Perkins, S. (2013), "Better Regulation of Public-Private Partnerships for Transport Infrastructure: Summary and Conclusions", *International Transport Forum Discussion Papers*, No. 2013/06, OECD Publishing, Paris. DOI: http://dx.doi.org/10.1787/5k46n41s7347-en.

Norwegian Ministry of Finance (2012), "Cost-benefit analysis", Official Norwegian Reports NOU (2012), October, Norway.

Pindyck, R.S. (2000), "Irreversibilities and the Timing of Environmental Policy", *Resource and Energy Economics*, Vol. 22, pp. 233-259.

Pindyck, R.S. (2013), "The climate policy dilemma", *Review of Environmental Economics and Policy, Association of Environmental and Resource Economists*, Vol. 7(2), pp. 219-237, July.

Pindyck, R.S. (2011), "Fat-tails, thin tails, and climate change policy", *Review of Environmental Economics and Policy*, Vol. 5(2), pp. 258-74, Summer.

Pires, F, Assis, L and Fiho, M (2012), "A real options approach to ship investment appraisal", *Journal of Business Management*, Vol. 6(25), pp.7397-7402, June.

Quinet E. (2013), "Factoring sustainable development into project appraisal: a French view", *ITF Discussion Paper*, No. 2013-31, October.

Trigeorgis, L. (1996), *Real options, managerial flexibility and strategy in resource allocation*, MIT Press, Cambridge, MA.

Weitzman, M.L. (2009), "On modelling and interpreting the economics of catastrophic climate change", *Review of Economics and Statistics*, Vol. 91(1), pp. 1-19.

Chapter 5

Discounting long-term effects of climate change for transport

This chapter looks at the theories and approaches to establish a discount rate for assessing long-term projects, discounting under risk and uncertainty and a comparison of how various countries have applied discount rates for climate change projects and policies.

The importance of the discount rate

Discounting is an integral part of any analysis such as cost-benefit analysis (CBA) that considers the costs and benefits over a number of years. Its aim is to express all costs and benefits in terms of their present value by assigning smaller weights to those that occur further in the future than to those that occur more immediately.

The CBA of long-term projects is particularly sensitive to the choice of the discount rate. For example, at an annual discount rate of 3%, the present value of $1 000 in 30 years' time is $412, compared to $231 at 5% and $742 at 1% (Figure 5.1). In 100 years' time, the present value of $1 000 reduces to $52 (at 3%), $8 (at 5%) and $370 (at 1%).

Figure 5.1. **Present value (of $1 000) varies by discount rate and time**

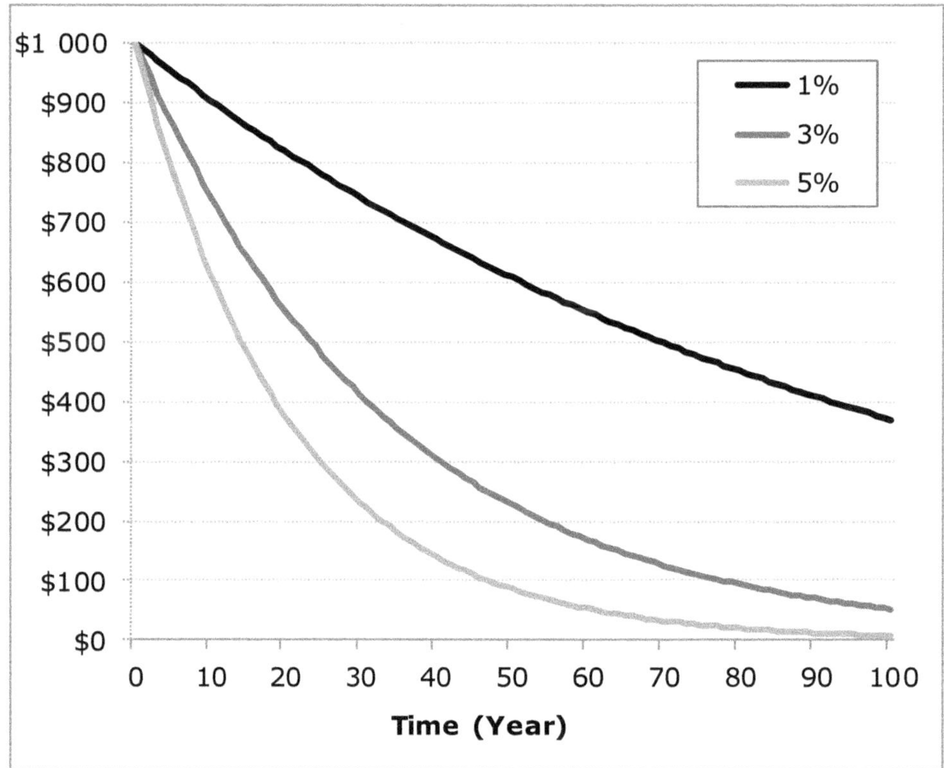

"While everyone agrees that the choice of discount rate is a crucial determinant of the value of public projects, there is less agreement on the appropriate discount rate to use to calculate present value. Academics, cost-benefit guides and textbooks give widely conflicting advice." (Harrison, 2010)

Discount rates and intergenerational concerns

Discounting can be adjusted to address intergenerational problems, which are often emphasised in the climate context. In tackling climate change there is a perceived need for the current generation to sacrifice their well-being in order to preserve the well-being of future generations. There are a number of ways that the current generation can protect the welfare of future generations, including leaving them with physical capital stock or better environmental stock (Harrison, 2010).

The choice of discount rate reflects the level of altruism the current generation has towards future generations. A higher discount rate ascribes future benefits lower weightings. With a high discount rate, few climate policies would pass the CBA test, resulting in less investment to protect future generations from global warming. But a low discount rate can sometimes also encourage counter-productive policies or projects from a climate policy perspective (OECD, 2007). For example, a low discount rate can encourage investment in long-lived coal-fired power stations with low operating costs but long pay-back periods for recovering capital investment instead of investment in gas-fired plants that have the opposite characteristics; high operating costs but a shorter pay-back period. Using a low discount rate also means that the current generation could invest in low-return projects at the expense of investments with higher return and thus make future generation worse off (Harrison, 2010).

There are two main approaches to determine discount rates for projects affecting future generations. They are the "prescriptive" and "descriptive" approaches to discount rate selection (Arrow et al., 1996; Harrison, 2010; Arrow et al., 2013a).

The prescriptive approach directly specifies a discount rate or parameters used in estimating the discount rate based on ethical principles or policy choices. Where the "prescriptive" approach to setting the discount rate is chosen, setting a high discount rate or even a flat discount rate could be seen as "unethical" (e.g. Ramsey, 1928). Under this approach, the social pure time preference becomes a "policy parameter" (Pindyck, 2013), which balances the welfare of current and future generations. If both generations are to be treated equally, the social rate of pure time preference should be lower (or zero), implying a lower discount rate. If the current generation is to be given more weight than the future generation, the rate of pure time preference increases, leading to a higher discount rate.

The descriptive approach, on the other hand, sets the discount rate based on observation of market behaviour (Pearce and Ulph, 1999; OECD, 2007; Kane, 2012). For example, the parameters in the Ramsey formula can be inferred by using empirical evidence to estimate the population's rate of time preference. Proponents of the descriptive approach suggest that the discount rate should approximate the market interest rates for long-term financial assets (such as government bonds) (Barro and Becker, 1989; Harrison, 2010; Kane, 2012). However, "market rates are conceptually distinct from a social discount rate" and "only reflect the preferences of current individuals, about their current decisions" and "not the interests of future individuals nor the preferences of current individuals about intergenerational matters" (OECD, 2007).

In a traditional transport appraisal framework, the discount rate is often assumed to be constant over time. Having a constant discount rate means individuals are time-consistent and that their later preferences confirm earlier preferences (Frederick et al., 2002). The theory of a declining discount rate was first developed by Weitzman (1998) and subsequently by Gollier and Weitzman (2010) and Freeman (2010). A key conclusion from those studies is that when future discount rates are uncertain, then the "effective" (or certainty-equivalent) discount rate must decline over time towards its lowest possible value (Gollier and Weitzman, 2010; Freeman, 2010 and USG, 2010).

Empirical literature seems to conform to the theory that discount rates are not constant over time (OECD, 2007 or Frederick et al., 2002). In their literature review, Frederick et al. (2002) found some empirical regularities regarding to discount rate including: asymmetric preference between gains and losses (gains are discounted more); small amounts are discounted more than large amounts and people seem to have a preference for spreading consumption over time. In addition, results from experiments[1] suggest that the discount function at the individual level declines over time (OECD, 2007).

Typical arguments for a declining discount rate include: falling economic growth rates, the uncertainty associated with future growth in per capita consumption and economic conditions, shocks to consumption due to catastrophic risks, changes to or heterogeneity in future preferences and intergenerational equity (OECD, 2007; Gollier and Weitzman, 2010; Arrow et al., 2014).

To illustrate the effects of discount rate on real SCC, Figure 5.2 provides a stylised illustration of the effect where real SCC grows at 5% per annum. Assuming the real SCC in year 0 is \$50 per tonne of CO_2, it increases to around \$6 600 after 100 years (before discounting).

If these values were discounted at a constant 5% per annum (i.e. same as the rate of increase in real SCC), the present value of real SCC will remain unchanged over time (at \$50 in this example). On the other hand, if the estimates were discounted at a declining discount rate (e.g. from 5% reducing to 3.7%), the real SCC in present value after 100 years would be much higher. In this stylised example, it is around \$175 per tonne of CO_2.

Figure 5.2. **Stylised interpretation of the effect of discounting on carbon value (t/CO_2)**

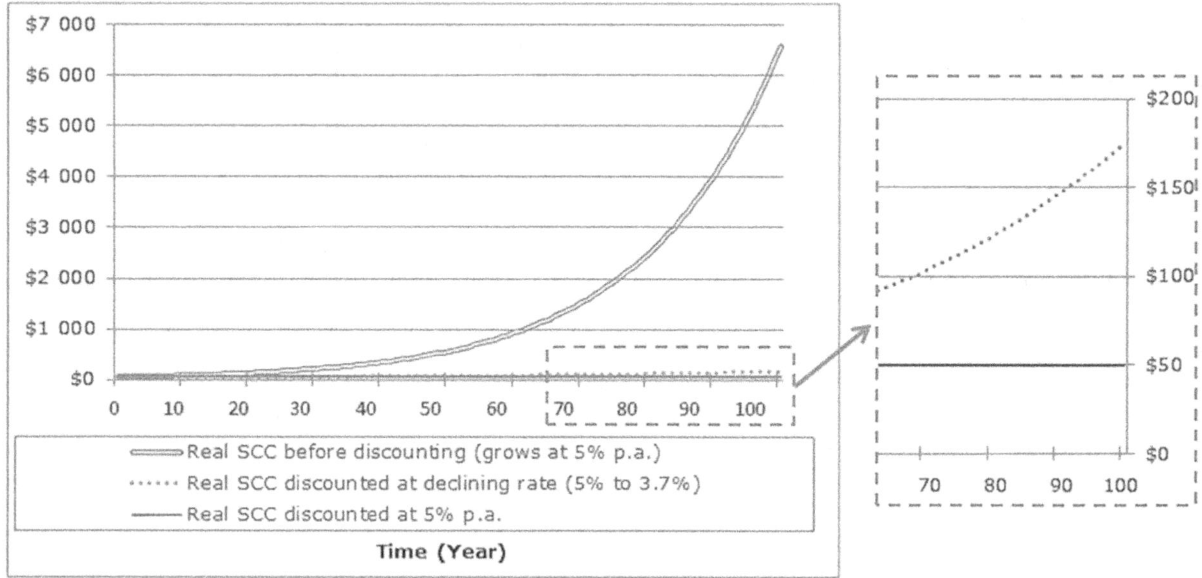

Discount rates for long-term projects

Discounting is a means for assessing outcomes over time by reference to individual, market or social preferences (especially for decisions that affect a long time-horizon). There are two commonly cited arguments for why this is necessary – positive time preference and the opportunity cost of investment (Harrison, 2010). These two arguments vary by the assumption as to whether private consumption or private investment will be displaced by public investment decisions. The marginal social cost of capital approach[2] based on the Capital Asset Pricing Model framework is a common approach to establish a discount rate to account for the displacement of private investment. This approach has been used by some countries (e.g. New Zealand and Japan) to determine the public sector discount rate. However, the pure time preference and the displacement of private consumption approach have received most attention in the current social discount rate literature due to its relevance to the assessment of the welfare of future generations (e.g. Weitzman, 2012 and Armtiage, 2014). The following chapters briefly outline three such approaches.

Ramsey formula

The positive time preference argument asserts that most individuals have a "pure preference" for the present, and also expect that as incomes increase over time the marginal utility of consumption declines and therefore they would prefer to consume now than to consume in the future. Otherwise, they would have to be compensated (e.g. through interest on savings) for delaying the consumption until the future.

The approach used to determine the discount rate under the positive time preference argument is the **social rate of time preference (SRTP)**. This approach reflects the impact of savings and investment on domestic consumption and the time preference individuals have on consumption today over the same level of consumption at a later date. It suggests the correct discount rate should be the rate at which a society is willing to postpone current consumption in exchange for future consumption without any change in overall wellbeing.

Box 5.1. **Ramsey formula**

The Ramsey formula, which has led academic research since the 1920s (Ramsey 1928) defines the discount rate (ρ_t) as follows:

$$\rho_t = \delta + \eta\, g_t.$$

δ	represents pure time preferences, which reflects individuals' preference for consumption now rather than in the future
η	the absolute value of the elasticity of marginal utility of consumption
g_t	the expected growth rate in per capita consumption between now and time t.
$\eta\, g_t$	represents the wealth effect related to the idea that future generations will be better off compared to present generations.

Although literature has suggested using the after-tax rate of return of low-risk marketable securities (such as government bonds) to approximate SRTP, the commonly used approach is the formula developed by Ramsey in 1928 (Box 5.1). The Ramsey formula has two key components, the pure time preference (δ) and the diminishing marginal utility of consumption over time ($\eta\, g_t$). The elasticity of marginal utility of consumption (η) represents the curvature of the utility function, a measure of aversion to interpersonal inequality and a measure of personal risk aversion[3] (Weitzman, 2007).

In the Ramsey formula, the discount rate is expressed as a function of expected growth rate in per capita consumption, therefore the resulting discount rate is not constant over time. If future consumption growth will be positively correlated with economic cycles (i.e. cyclical), the discount rate should vary with the economic cycles (OECD, 2007).

Extended Ramsey formula

The Ramsey formula has often been used as a basis for intergenerational discounting by including an extra term to account for the precautionary effect around future rate of growth in consumption (Box 5.2) (Gollier, 2002; Weitzman, 2007; OECD, 2007; Arrow et al., 2013b and Cropper et al., 2014). According to this formula, the precautionary effect gets bigger as the variance of future consumption increases and therefore results in a lower discount rate.

Box 5.2. **Extended Ramsey formula to account for precautionary effect** (note)

The modified Ramsey formula is given by:

$$\rho_t = \delta + \eta\, g_t - \tfrac{1}{2}\eta^2 \sigma_g^2$$

where:

ρ_t, δ, η and g_t	are defined as in Box 5.1.
σ_g^2	variance of consumption at time t, and
$\tfrac{1}{2}\eta^2 \sigma_g^2$	a precautionary effect.

Note: This formula assumes growth rate in consumption is independently and identically distributed over time (i.e. it follows a random walk or arithmetic Brownian motion).

Systemic risk-adjusted Ramsey formula

Although there has been no consensus on the discount rate for public sector appraisal, there is some consensus on the need to account for the uncertainty associated with the linkages between future project benefits and future macroeconomic conditions (e.g. Weitzman, 2007 & 2012; Quinet, 2013; Gollier, 2013 & 2014). Related literature mentions another conceptual version of the Ramsey formula, the systemic risk-adjusted SRTP[4]. This approach is similar to the extended Ramsey formula with an extra term that links project benefits and costs with Gross Domestic Product (GDP).

In this approach, the discount rate is expressed as the sum of the risk-free interest rate plus the product of the risk premium[5] (φ) and the correlation between project benefits and economic activity (β) (Box 5.3). The risk-free interest rate is the same as the extended Ramsey formula (Box 5.2). This systemic risk-adjusted Ramsey formula is referred as the consumption-based CAPM by Gollier (2014) because of its similarity to the standard CAPM[6]. Gollier (2014) shows that systemic risk premium increase with uncertainty over time and therefore the risk-adjusted discount rate can increase over time if the beta is higher than $\eta/2$.

Box 5.3. **Systemic risk-adjusted Ramsey formula** (note)

The systemic risk-adjusted social rate of time preference is given by:

$r = r_f + \varphi \beta$

where:

r is the risk-factored discount rate specific to the project

r_f is the risk-free rate (i.e. the extended Ramsey formula);

$r_f = \delta + \eta g_t - \frac{1}{2} \eta^2 \sigma_g^2$

φ is the general risk premium, a parameter common to all projects, that measures the amplitude of the long-term systemic risks linked to macro-economic trends;

$\varphi = \eta \sigma_g^2$

β is beta, a project-specific parameter that measures the correlation between project benefits and economic activity.

Note: This formula assumes the evolution of economic activity is independently and identically distributed over time (i.e. it follows a random walk or arithmetic Brownian motion).

The key rationale of this approach is that each project entails various types of risk, including those that are associated with the future overall macroeconomic conditions (i.e. φ). If the project's benefits are positively correlated with the macro-economic conditions (i.e. β is positive), the risk on project returns gets amplified, particularly in the case of large-scale transport projects in which the returns would be unexpectedly lower under bad macro-economic outcomes.

For transport projects, key benefits (such as time savings) are typically subject to uncertainty. There is also a risk around unexpected change in travel demand. Facing this risk, investors (including governments) would be more cautious in their decision making, leading them to increase the discount rate: whenever future travel demand is heavily affected by the macroeconomic conditions, the distribution of future outcomes becomes more spread, hence inducing a positive risk premium.

For climate policy, on the other hand, the issue is more complicated as the correlation between the returns from climate policy and the risk at the macro-economic level is unclear. As the economy grows, more activities, including transport, produce more GHG. This implies a positive correlation between GHG emissions and the macroeconomic conditions. However, emissions may also have a negative impact on economic growth, as climate impacts may cause significant damages to the economy. Due to the presence of this feedback effect, the overall effects are ambiguous.

One benefit of this systemic risk-adjusted SRTP is that it allows calculation of the discount rate project by project (Quinet, 2013). Since the discount rate is determined by three factors (the risk-free rate, the beta correlation between projects and the economy, and the risk premium), it can be constant or it can vary over time-period depending on the values of the parameters chosen.

Discounting under risk and uncertainty

There are two broad categories of uncertainty that affect the choice of discount rate. They are the uncertainty around future interest rates and/or the components of the social discount rate (such as growth) and the uncertainty around future benefits[7] due to project risks.

Uncertainty without project risk

Without project risks, a risk neutral social planner will adopt a discount rate close to the risk-free rate (based on the extended Ramsey formula). When the discount rate is unknown, the literature suggests using different choices of discount rate to derive the certainty equivalent (CE) discount rate (e.g. Weitzman, 1998 and Gollier and Weitzman, 2010). The key argument of this approach is that what should be probability-averaged are not the future discount rates at various time periods but the future discount factors (Weitzman, 1998; Freeman, 2010 and Traeger, 2013). Discount factors are the factors by which future cash flows must be multiplied to obtain the present value, i.e., if the discount rate is *d*, the discount factor for year *i* is represented by $\frac{1}{(1+d)^i}$.

To illustrate the approach, Table 5.1 shows how discount rates can be combined. The average discount factors for 3% and 7% discount rates are higher than the discount factors for a 5% discount rate (i.e. average of 3% and 7%). The implicit discount rate derived from the average discount factors is called the certainty-equivalent (CE) discount rate (Weitzman, 1998). In this example, the CE discount rate declines over time and approaches the low-end of the discount rate range over time. Thus, the risk-free rate may decline over time due to uncertainty.

Table 5.1. **Numerical example of a declining certainty-equivalent discount rate**

Year	Discount factor for a 3% rate	Discount factor for a 7% rate	Certainty-equivalent discount factor (average) - note	Certainty-equivalent discount rate
1	0.971	0.935	0.953	5.0%
10	0.744	0.508	0.626	4.8%
20	0.554	0.258	0.406	4.6%
30	0.412	0.131	0.272	4.4%
40	0.307	0.067	0.187	4.3%
50	0.228	0.034	0.131	4.1%
60	0.170	0.017	0.093	4.0%
70	0.126	0.009	0.068	3.9%
80	0.094	0.004	0.049	3.8%
90	0.070	0.002	0.036	3.8%
100	0.052	0.001	0.027	3.7%
200	0.003	1.33E-06	0.001	3.4%
300	1.41E-04	1.53E-09	7.04E-05	3.2%
400	7.33E-06	1.76E-12	3.67E-06	3.2%

Note: This example assumes the probability for the two discount rates to occur is the same. If the estimates of probability for different discount rates (could be more than two) are available, the probability-weighted average should be used. These probability-weights can differ between time periods (Gollier and Weitzman, 2010).

The CE approach is one way to gauge what the average discount rate would be at different points in time. However, for the CE approach to be valid, it is necessary for the discount rate to be persistent (i.e. period of low or high will tend to be followed by further periods of low or high rates). Literature has found evidence to support this persistency in interest rate (e.g. OECD, 2007; Groom et al., 2007; Freeman et al., 2013).

Uncertainty with project risk

According to Arrow and Lind (1970) the total risk of public investment can be shared between a large number of individuals and therefore the risk burden to individuals for inclusion in CBA becomes negligible. Due to transaction costs and market imperfections, however, the risk premium for a public investment is not zero. It has been suggested (e.g. Sandmo, 1972; Weitzman, 2012; Quinet, 2013 and Gollier, 2014) that the public sector's discount rates should include a risk premium. With project risks, project benefits become uncertain. In theory, the risk premium is likely to increase with uncertainty. As noted, while the risk-free rate may decline over time due to uncertainty, the risk premium is likely to increase over time. Therefore, the systemic risk-adjusted discount rate (e.g. using the systemic risk-adjusted Ramsey formula) can increase or decrease over time, depending on the relative force of the two effects (Gollier, 2014).

Risk and uncertainty

To account for the preceding treatment of uncertainty with and without project risk in CBA, a common approach would be to apply objective probability distributions (of risk) to economic growth and project returns, taking account of correlations. Theoretically speaking, however, such an approach only considers risk but not Knightian uncertainty. As distinguished by Knight (1921), measureable uncertainty (i.e. risk) is "so far different from an unmeasured one that it is not in effect an uncertainty at all". Since uncertainty is not measureable, it is simply not possible to assign a probability or statistical distribution to estimate the expected outcomes. In practice though, the magnitude of the macro-economic risk premium captures a certain degree of uncertainty. This risk premium may be supported by probability distributions of growth scenarios, and in Quinet (2013) by a more general subjective description of the magnitude of the uncertain macro-economic risk. In the latter case, risk and uncertainty are mingled together and their combined consequences are captured to a certain degree, which overstates low risk and small Knightian uncertainty but understates extreme risks and high Knightian uncertainty.

In recent literature (e.g. Klibanoff et al., 2005 and Traeger, 2014), there are models that attempt to examine how ambiguity (one of the multiple forms of uncertainty) affects the discount rate. These models apply a subjective probability distribution over objective probability distributions to capture the uncertainty about the correct objective probability distribution (Traeger, 2014). Results show that "a decision-maker who is more averse to ambiguity than to risk will lower the discount rate more for [ambiguity] than for [risk]" (Traeger, 2014). As the wide area of research currently being developed beyond the classical expected utility maximising framework produces results and improves over time, practical steps to account for some aspect of Knightian uncertainty may become possible.

International comparison

This section briefly summarises discount rate practices from the partial survey of OECD member countries. Box 5.4 provides more details on each country's approach.

Currently, different countries apply different discount rates in CBAs (Table 5.2). The marginal social opportunity cost of capital and the social rate of time preference are the two key approaches used by most jurisdictions to estimate discount rates. The former methodology tends to result in a higher discount rate. Differences in preferences, term structure of interest rate, correlation between projects and economic conditions also contribute to the observed differences in the discount rates chosen.

The United Kingdom and Norway adjust the discount rate[8] for the risk associated with long-term effects by adopting a declining schedule. The Netherlands, Germany and the United States instead adopt a lower but constant discount rate.

Table 5.2. **Transport sector discount rate in different countries**

Country	Method	Discount rate		
France	Risk-adjusted SRTP	Constant: 4.5% or project specific rate		
The Netherlands	Risk-adjusted SRTP	4% for climate change effects and 5.5% for other effects		
Norway	Risk-adjusted SRTP	<40 years: 4%	40-75 years: 3%	>75 years: 2%
UK	SRTP	0-30 years: 3.5%	31-75 years: 3%	Reducing to 1% for over 300 years
Sweden	SRTP	Constant 3.5%		
Germany	SRTP	Constant 1% for long-term climate change effects, 1.5% for other effects and 3% for short term effects (0-20 years)		
US	Certainty equivalent	Constant: 2.5%, 3%, and 5% (for estimation of SCC)		
Japan	SOC	Constant 4%		
New Zealand	SOC	8% as recommended by NZ Treasury (6% used by NZ Transport Agency)		

Note: SOC – Marginal social cost of capital; SRTP – Social Rate of Time Preference (based on variants of the Ramsey formula).

Source: A preliminary OECD survey of carbon values in selected countries (Chapter 5).

To illustrate the impact of risk-adjusted discounting on the final carbon value used in CBA, the carbon values used by France, The Netherlands and Norway are discounted first using the risk-free component of the risk-adjusted discount rate and then by the additional risk-premium (i.e. the risk-adjusted discount rate) (Figure 5.3). The effects of risk-adjustment are not insignificant.

Figure 5.3. **Carbon value with risk-adjusted discounting for selected countries (in USD 2013 values/tCO_2)**

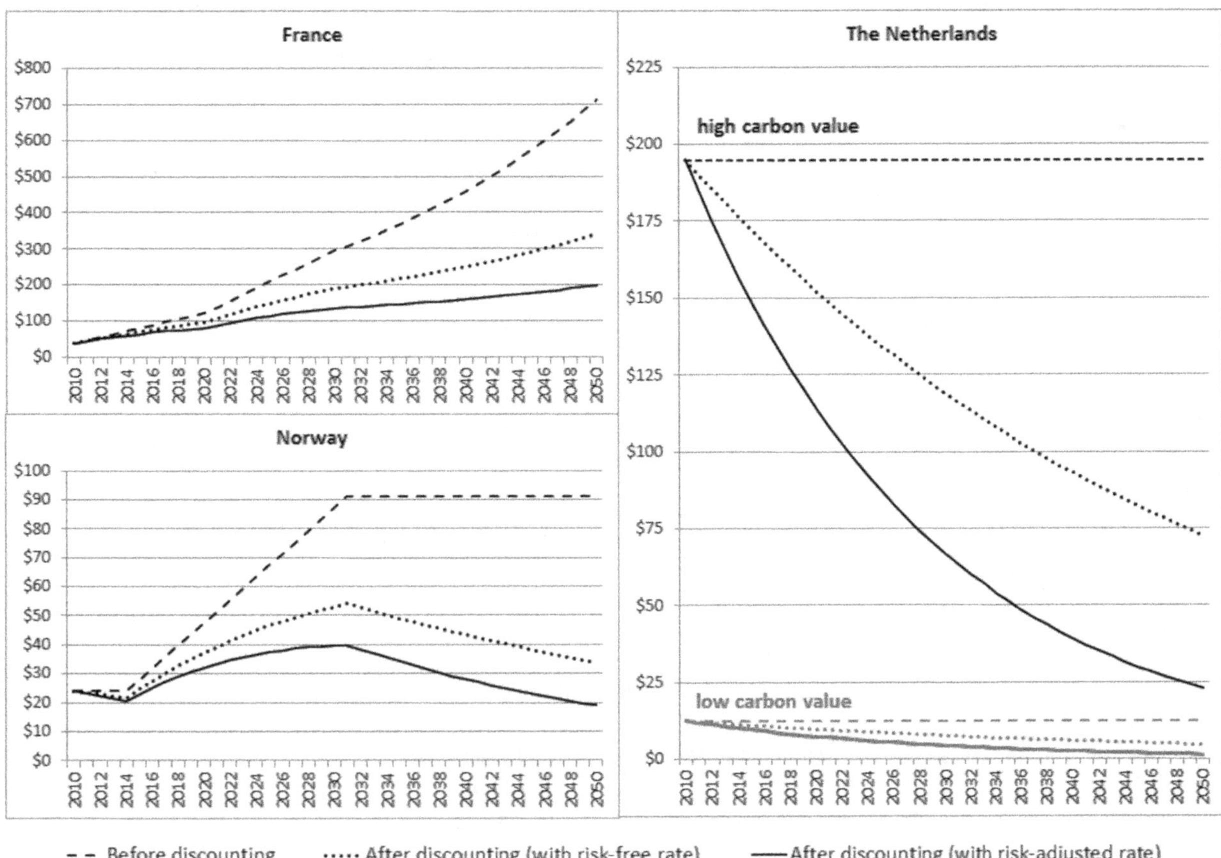

Source: ITF calculations based on OECD survey of selected member countries.

Box 5.4. Discount rates: International practices

United States

Government agencies in the United States traditionally used constant discount rates of 3% and 7% in their CBAs. However, after considering intergenerational issue, the interagency group ultimately chose three certainty-equivalent constant discount rates: 2.5%, 3%, and 5% per year (Greenstone et al., 2013).

France

In France, the discount rate is based on the systemic risk-adjusted SRTP approach to take into account systemic risk with a risk-free rate of 2.5% (falling to 1.5% from 2070) and a risk premium of 2% (rising to 3% from 2070) multiplied by a sector specific (or, when available, a project specific) beta value (Quinet, 2013). The increasing risk premium reflects the increase in project-specific systemic risks as the time horizon extends. With a beta of 1, the standard discount rate is 4.5%. However, the discount rate can vary between projects as the beta value varies.

United Kingdom

The UK uses Social Rate of Time Preference to set the discount rate. Because of the uncertainty about the future values of time preference, a certainty equivalent rate taking into account the range of this uncertainty was calculated (HM Treasury, 2011). In the end, the UK Green Book recommends the following discount rate for different time periods (Table 5.3).

Table 5.3. **Green Book discount rates**

Years from current year	Discount rate
0-30	3.5%
31-75	3.0%
76-125	2.5%
126-200	2.0%
201-300	1.5%
301 and over	1.0%

Source: HM Treasury (2011).

Norway

In Norway, the discount rate has two components – a risk-free rate and a risk premium. For evaluation periods under 40 years, Norway uses a risk-adjusted rate of 4%, which is the sum of a 2.5% risk-free rate and a 1.5% risk premium. Based on a declining risk-free rate and a declining risk premium, the discount rate in Norway reduces to 3% for years from 40 to 75 and to 2% from year 75 onwards (Table 5.4).

Table 5.4. **Discount rate in Norway**

Discount rate	Years 0 -40	Years 40 -75	From year 75
Risk-free rate	2.5%	2%	2%
Risk premium	1.5%	1%	0%
Risk-adjusted rate	**4.0%**	**3%**	**2%**

Source: Norwegian Ministry of Finance (2012: 78-79).

Box 5.4. **Discount rates: International practices** *(cont.)*

Japan

Japan uses the SOC approach to determine discount rate for CBA. The current discount rate is set at 4%, and it is constant throughout the evaluation period (maximum 50 years). More details of Japan's approach are provided in Annex A.

Netherlands

The discount rate for the Dutch government CBA has two components – a fixed risk-free rate based on information obtained from the capital market (of 2.5%) and a risk premium (of 3%) to account for relative risk in the future. Therefore the discount rate is set at 5.5%. However, for external effects that are irreversible (such as climate change), the risk premium is 1.5% (instead of 3%), giving a discount rate for climate change effects of 4%. More details of the Netherlands' approach are provided in Annex B.

Sweden

Sweden uses social rate of time preference (SRTP) and the Ramsey formula to set the social discount rate. After considering the arguments for a declining discount rate due to increasing risk over time, Sweden chose to set a discount rate at a lower level to approximate a declining schedule with a single average figure. The discount rate in Sweden is currently set at 3.5%.

Germany

The standard social discount rate (based on SRTP) in Germany for cross-generational valuations is 1.5%. For long-term climate change effects, UBA recommends a constant discount rate of 1%. This corresponds to a more conservative estimate of an annual economic growth rate of 1% over the next 100 years. In Germany, the constant discount rate applies to the entire evaluation period. More details of Germany's approach are provided in Annex C.

New Zealand

Prior to 2008, the public sector discount rate used in New Zealand was set at 10%. In 2008, following a review of discount rate methodologies and parameters for establishing the discount rate, the NZ Treasury recommended an 8% discount rate (after tax, real) for use in transport investment decisions. This estimate was based on the social opportunity cost of capital (SOC) approach and used industry data to estimate the parameters. Following the 2008 update, NZ Transport Agency also amended the discount rate to 8% in the same year. In 2013, NZ Transport Agency reviewed the parameters used in the SOC formula and decided to use a 6% discount rate instead.

At present, transport infrastructure projects that are funded by NZ Transport Agency are assessed using a 6% discount rate (and a 40-year evaluation period). However, for investment decisions that require Crown funding (e.g., the decision on whether to build a new ferry terminal at Clifford Bay to replace an existing ferry terminal at Picton) and for policy decisions (e.g. whether to reduce the adult legal blood alcohol concentration limit), the NZ Ministry of Transport adopts NZ Treasury's 8% discount rate.

Source: OECD survey of selected member countries.

Notes

1. The shape of the discount function can be constructed by asking people to choose between a set of delayed rewards, such as money and sweets (OECD, 2007).

2. The marginal social cost of capital (SOC) can be simplified into three key terms: the after-tax real risk-free interest rate (r^f), an asset beta (β) and a (tax-adjusted) risk premium (R^p). The formula is given by: $r = r^f + \beta\ R^p$ (see for example: Spackman, 2008; NZ Treasury, 2008; Weitzman, 2012 and Armitage, 2014). A major criticism of the SOC approach is the lack of a logical mechanism to derive the risk premium for a public project as the standard approach is based largely on the financial markets (Spackman, 2008). Furthermore, SOC does not consider the interest of future generations and how the current generation sees intergenerational matters (OECD, 2007 and Armitage, 2014).

3. There are three categories of risk preference: risk aversion, risk-neutral and risk-taking. The utility function under a risk aversion assumption is concave, whereas it is linear and convex under the risk-neutral and risk-taking assumptions respectively.

4. By considering the correlation between the increased output of the project and returns to the economy as a whole, Weitzman (2007) demonstrated that discount rate at time t can be expressed as $r_t = \beta\ r_e + (1\text{-}\beta)\ r_f$ (where r_e is the expected return from investment for the economy as a whole and r_f is the risk-free rate). By substituting the standard risk premium expression $r_e - r_f = \eta\sigma_g^2$, the discount rate equation reduces to $r_t = r_f + \beta\ \eta\sigma_g^2$. One result of Weitzman's discount rate equation is that the discount rate declines monotonically over time to approach the risk-free rate (Weitzman, 2007 p.711 and Weitzman, 2012 p.25).

5. In this case, the risk premium measures the amplitude of the long-term systemic risks linked to macro-economic trends.

6. Capital Asset Pricing Model.

7. In transport appraisal, uncertainty occurs in many areas, such as the level of emissions due to changes in technological advancement and/or modal changes, which will affect the expected project benefits.

8. Term structure of interest rate (also known as the yield curve) is financial term that describes the relationship between interest rates and time to maturity (known as the "term").

References

Armitage, S. (2014), “Discount rates for long-term projects: the cost of capital and social discount rate compared”, September 2014, The University of Edinburgh Business School, http://papers.ssrn.com/sol3/Delivery.cfm/SSRN_ID2505150_code102554.pdf?abstractid=2505150&mirid=1

Arrow, K.J. and Lind R.C. (1970), “Uncertainty and the evaluation of public investment decisions”, *The American Economic Review*, Vol. 60(3), Jun, pp.364-378.

Arrow, K.J. et al. (1996), “Intertemporal Equity, Discounting, and Economic Efficiency.” In *Climate Change 1995: Economic and Social Dimensions of Climate Change*, Contribution of Working Group III to the Second Assessment Report of the Intergovernmental Panel on Climate Change, edited by J.P. Bruce, H. Lee, and E.F. Haites. Cambridge: Cambridge University Press.

Arrow, K.J. et al. (2014), “Should governments use a declining discount rate in project analysis”, *Review of Environmental Economics and Policy*, Advance access July 2014. http://reep.oxfordjournals.org/content/early/2014/07/16/reep.reu008.full.pdf .

Arrow, K.J. et al. (2013a), “Determining benefits and costs for future generations”, *Science*, Vol. 341(6144), 349-350.

Arrow, K.J. et al. (2013b), “How should benefits and costs be discounted in an intergenerational context?”, *Working Paper Series*, 5613, Department of Economics, University of Sussex.

Barro, R. and Becker, G. (1989), “Fertility choice in a model of economic growth”, *Econometrica*, Vol. 57(2), March, pp. 481-501.

Cropper, M., Freeman, M., Groom, B., and Pizer, W. (2014), "Declining discount rates", *American Economic Review*, Vol. 104(5), pp. 538-43.

Frederick, S., Loewenstein, G. and O’Donoghue, T. (2002), “Time discounting and time preference: A critical review”, Journal of Economic Literature, Vol. XL, June, pp.351-401.

Freeman, M. (2010), “Yes, we should discount the far-distant future at its lowest possible rate: A resolution of the Weitzman-Gollier Puzzle”, *Economics: The Open-Access, Open-Assessment E-Journal*, Vol. 4, 2010-13, April.

Freeman, M. et al. (2013), “Declining discount rates and the Fisher Effect: inflated past, discounted future?”, Centre for Climate Change Economics and Policy and Grantham Research Institute on Climate Change and the Environment, http://www.lse.ac.uk/GranthamInstitute/wp-content/uploads/2014/02/WP109-discount-rates-fisher-effect.pdf

Gollier, C. (2014), “Discounting and growth”, *American Economic Review*, Vol. 104(5), pp. 534-37.

Gollier, C (2013), *Pricing the future: The economics of discounting in an Uncertain World*, Princeton University Press.

Gollier, C (2010), "Expected net present value, expected future value, and the Ramsey rule", *Journal of Environmental Economics and Management*, Vol. 59(2), March, pp. 142–148.

Gollier, C (2002), "Time horizon and the discount rate", *Journal of Economic Theory*, Vol. 107, pp. 463-473.

Gollier, C. and Weitzman, M. (2010), "How should the distant future be discounted when discount rates are uncertain?", *Economics Letters*, Vol. 107, pp.350-353.

Groom B. et al. (2007). "Discounting the Distant Future: How Much Does Model Selection Affect the Certainty Equivalent Rate?", *Journal of Applied Econometrics*, Vol. 22, pp. 641-656.

Harrison, M. (2010), "Valuing the future: the social discount rate in cost-benefit analysis", Visiting Research Paper, Productivity Commission, Australian Government.

HM Treasury (2011), "The Green Book: Appraisal and evaluation in central government", London, United Kingdom.

Kane, M. (2012), "Disagreement and design: An arbitration of the climate change and intergenerational discounting debate", New York University Law and Economics Working Papers, Paper 306 http://lsr.nellco.org/nyu_lewp/306.

Klibanoff, P., Marinacci, M. and Mukerji, S. (2005), "A smooth model of decision making under ambiguity", *Econometrica*, Vol. 73(6), pp.1849-1892.

Knight, F. (1921), *Risk, uncertainty and profit*, Boston, MA: Hart, Schaffner & Marx; Hougton Mifflin Company.

New Zealand Treasury (2008), "Public sector discount rates for cost benefit analysis", New Zealand Treasury, 2008.

Norwegian Ministry of Finance (2012), "Cost-benefit analysis", Official Norwegian Reports NOU (2012), October, Norway.

OECD (2007), "Use of discount rate in the estimation of costs of inaction with respect to selected environmental concerns", Working Party on National Environmental Policies, OECD, Paris, drafted by Hepburn, C., www.oecd.org/officialdocuments/publicdisplaydocumentpdf/?doclanguage=en&cote=env/epoc/wpnep(2006)13/final.

Pearce, D.W. and Ulph. D, (1999), "A social discount rate for the United Kingdom", in D. W. Pearce, *Environmental Economics: Essays in Ecological Economics and Sustainable Development*, pp. 268–285, Edward Elgar, Cheltenham.

Pindyck, R.S. (2013), “The climate policy dilemma”, *Review of Environmental Economics and Policy, Association of Environmental and Resource Economists*, Vol. 7(2), pp. 219-237, July.

Quinet, E. (2013), "Factoring Sustainable Development into Project Appraisal: A French View", *International Transport Forum Discussion Papers*, No. 2013/31, OECD Publishing, Paris. DOI: http://dx.doi.org/10.1787/5jz40rncm222-en

Ramsey, Frank P. (1928), "A mathematical theory of saving", *Economic Journal*, Vol. 38(152), pp. 543–559.

Sandmo, A. (1972), “Discount Rates for Public Investment Under Uncertainty”, *International Economic Review*, Vol. 13(2), June, pp. 287–302.

Spackman, M. (2008), “Time preference, the cost of capital and PPPs”, Conference on ‘Discount rates for the evaluation of Public Private Partnerships’, Queen’s University, Ontario, http://jdi-legacy.econ.queensu.ca/Files/Conferences/PPPpapers/Spackman-081002-final.pdf

Traeger, C.P. (2014), “Why uncertainty matters: discounting under inter-temporal risk aversion and ambiguity”, *Economic Theory*, Vol. 56(3), pp.627-664.

Traeger, C.P. (2013), “Discounting under uncertainty: disentangling the Weitzman and the Gollier effect”, CUDARE working paper no. 1121, UC Berkeley. http://papers.ssrn.com/sol3/Delivery.cfm/SSRN_ID2270459_code1081099.pdf?abstractid=2270459&mirid=1

USG (2010), “Technical support document: Social cost of carbon for regulatory impact analysis - Under executive order 12866”, Interagency Working Group on Social Cost of Carbon, United States Government, February.

Weitzman, M.L. (2012), “On risk-adjusted discount rates for long-term public investments”, Department of Economics, Harvard University, http://nyttekost-web.sharepoint.com/Documents/120908_RiskAdjustedDiscountRate.pdf .

Weitzman, M.L. (2007), “A review of the Stern Review on the economics of climate change”, *Journal of Economic Literature*, Vol. XIV, September, pp.703-724.

Weitzman, M. L. (1998), “Why the far-distant future should be discounted at its lowest possible rate”, *Journal of Environmental Economics and Management*, Vol. 36(3), pp.201-208.

Further reading

Ackerman, F. and Bueno, R (2011), "Use of McKinsey abatement cost curves for climate economics modelling", Stockholm Environment Institute, Working paper WP-US-1101.

Delink, R., Lanzi, E., Chateau, J., Bosello, F., Parrado, R., and de Bruin, K. (2014), "Consequences of climate change damages for economic growth: a dynamic quantitative assessment", *OECD Economics Department Working paper*, No. 1135, OECD Publishing, Paris. DOI: http://dx.doi.org/10.1787/5jz2bxb8kmf3-en

DECC (2013b), "Updated short-term traded carbon values used for UK public policy appraisal", September, Department of Energy and Climate Change, United Kingdom.

Gollier, C (2010), "Expected net present value, expected future value, and the Ramsey rule", *Journal of Environmental Economics and Management*, Vol. 59(2), March, pp. 142–148.

Heal, G. and A. Millner (2013), "Scientific Ambiguity and Climate Policy", *Environmental Resource Economics*, Vol. 55, pp.21-46.

Hope, C., J. Anderson and P. Wenman (1993), "Policy analysis of the greenhouse effect: An application of the PAGE model", *Energy Policy*, Vol. 21(3), pp. 327-338.

IPCC (2013), "Climate Change 2013: The physical science basis - summary for policy makers", Contribution of Working Group I.

Kesicki, F. and Ekins P. (2012), "Marginal abatement cost curves: a call for caution", *Climate Policy*, 12(2), pp. 219-236.

Kimball, M. (1990), "Precautionary saving in the small and in the large", *Econometrica*, Vol.58, No. 1 (Jan 1990), pp.53-73.

Meunier, D. (2012), "Towards a sustainable development approach in transport assessment", *Procedia-Social and Behavioral Sciences*, 48, 3065-3077.

Meunier, D. and Quinet, E. (2014), "Valuing GHG emissions and uncertainty in transport cost benefit analysis", European Transport Conference 2014, Frankfurt, Germany, September.

Millner, A. (2013), "On welfare frameworks and catastrophic climate risks", *Journal of Environmental Economics and Management*, 2013, http://dx.doi.org/10.1016/j.jeem.2012.09.006

Nordhaus, W. D. (1992), "The 'DICE' Model: Background and Structure of a Dynamic integrated climate-economy model of the economics of global warming", Cowles Foundation Discussion Papers 1009, Cowles Foundation for Research in Economics, Yale University.

Nordhaus, W. D. (2009), "An analysis of the dismal theorem", *Cowles Foundation Discussion Paper*, No. 1686, Yale University.

Nordhaus, W. (2011), "The economics of tail events with an application to climate change", *Review of Environmental Economics and Policy*, Vol. 5(2), pp. 240–257.

OECD (2006), *Cost-Benefit Analysis and the Environment: Recent Developments*, OECD Publishing, Paris, http://dx.doi.org/10.1787/9789264010055-en.

Proost, S. and Van Dender, K. (2011), "What long-term road transport future? Trends and policy options", *Review of Environmental Economics and Policy*, Vol. 5(1), pp. 44-65.

Proost, S., E. Delhaye, W. Nijs, and D. Van Regemorter (2009), "Will a radical transport pricing reform jeopardize the ambitious EU climate change objectives?", *Energy Policy*, Special Issue, Vol. 37(10), pp. 3863-71.

Rienstra, S. (2008), "*De rol van kosten-batenanalyse in de besluitvorming*" (The role of cost-benefit analysis in decision making), Kennisinstituut voor Mobiliteitsbeleid.

Small, K. A. (1991), "Transportation and the environment," in Roland T. (Ed.), *The Future of Transportation and Communication*, Swedish National Road Administration, pp. 217-230.

Stern, F. et al. (1999), "Summary of experimental uncertainty assessment methodology with example", Iowa Institute of Hydraulic Research, College of Engineering, The University of Iowa, IIHR Technical Report No. 406 http://user.engineering.uiowa.edu/~cfd/pdfs/References/uncert.pdf.

Tol, R.S.J. (1996), "The climate framework for uncertainty, negotiation and distribution", in: K.A. Miller and R.K. Parkin (Eds.), *An Institute on the Economics of the Climate Resource*, University Corporation for Atmospheric Research, Boulder, pp. 471-496.

van Dender, K. and M. Clever (2013), "Recent Trends in Car Usage in Advanced Economies – Slower Growth Ahead?: Summary and Conclusions", *International Transport Forum Discussion Papers*, No. 2013/09, OECD Publishing, Paris. DOI: http://dx.doi.org/10.1787/5k46n4121k5d-en

Vielle, M. and Bernard A. L. (1998), "La structure du modèle GEMINI-E3", *Économie et Prévision*, Vol. 136(5), pp. 19-32.

Warner, J.T. and Pleeter, S. (2001), "The personal discount rate: evidence from military downsizing programs", *American Economic Review*, Vol. 91(1), pp. 33–53.

Young, L (2002), "Determining the discount rate for government projects", New Zealand Treasury, Working Paper, 02/21.

List of participants

Working group members

- Jean-Jacques Becker, Ministry of Ecology, Sustainable Development and Energy, France [Chair]
- Kilian Frey, Federal Environmental Agency, Germany
- Bjorn Bunger, Unweltbundesant, Germany
- Anco Hoen, PBL Netherlands Environmental Assessment Agency, the Netherlands
- Catharina Horn, Federal Ministry of Transport & Digital Infrastructure, Germany
- Hironori Kato, University of Tokyo, Japan
- Hans Nijland, PBL Netherlands Environmental Assessment Agency, the Netherlands
- Lars Nilsson, Swedish Transport Administration

External experts

- Emile Quinet, ENPC and Paris School of Economics, France [Rapporteur]
- Svante Mandell, VTI Swedish National Road and Transport Research Institute, Sweden [Rapporteur]
- Nils Axel Braathen, OECD Environment Directorate [presenter]
- Elizabeth Kopits, US Environmental Protection Agency [presenter]
- Bruno Faivre D'Arcier, Université de Lyon, France
- Kurt Van Dender, OECD Centre for Tax Policy
- Mark Freeman, Loughborough University, UK
- Ben Groom, London School of Economics, UK
- Marc Ivaldi, University of Toulouse, France
- David Meunier, Ministry of Ecology, Sustainable Development and Energy, France
- Antony Millner, London School of Economics, UK
- Ian Parry, International Monetary Fund
- Stephen Smith, University College London
- Nicolas Treich, University of Toulouse, France
- Cees Withagen, University of Amsterdam, Netherlands

Annex A

Carbon value and discount rates in Japan

Perspective of Professor Hironori Kato, the University of Tokyo, Japan

January 2015

Introduction

In recent years, public concerns about global warming have been growing in Japan[1]. In 2011, carbon dioxide (CO_2) emissions from the transport sector accounts for 17.9% in Japan, which represents an increase of 5.2% compared to the 1990 level (Ministry of the Environment, Japan (MOE), 2013). To reduce the level of CO_2 emissions from the transport sector, Ministry of Land, Infrastructure, Transport and Tourism, Japan (MLIT) has in recent years implemented a number of measures such as: development and promotion of environmental-friendly automobiles; promotion of smart car use, improvement of traffic conditions; improvement in freight transport efficiency, promotion of public transport use; and improving energy consumption efficiency in rail, ship and aviation transport (MLIT, 2012).

Transport related CO_2 emissions are highly dependent on the level of demand for transport that is fuelled by fossil energy. Transport infrastructure investment (e.g. public transport projects) can play a role to reduce the overall CO_2 emission level, which in turn should contribute to the achievement of a more sustainable transport system in the long run. Thus, incorporating CO_2 emission effects in the evaluation of transport projects is important.

This annex first summarises the current cost-benefit analysis (CBA) practice for evaluating transport infrastructure investment in Japan. It then discusses the valuation of the CO_2 emissions in the CBA of transport projects.

CBA practice in Japan

A formal CBA process for evaluating transport infrastructure investment was first introduced by the Government of Japan in the late 1990s. All publicly-funded transport projects are evaluated in accordance with the CBA Manuals published by MLIT (to be discussed below). The CBA estimates the benefits and costs to be expected from a transport project. These benefits and costs include user's benefit, supplier's profit, and external impacts (including CO_2 emissions) in monetary terms.

The CBA Manuals for various types (i.e. airport, rail, road and seaport) of transport infrastructure investment projects have been revised several times by MLIT since their initial development in the

1990s. These manuals were independently prepared by different bureaus under the MLIT to reflect the different characteristics of the transport facilities or services and were last updated between 2008 and 2012 (see Table A.1). Transport projects that are funded or subsidised by the national government are required to follow the CBA Manuals in their project evaluation.

Table A.1. **CBA Manuals in Japan**

Project type	Manual title	Latest updated
Airport	Cost-effectiveness Analysis Manual of Airport Development Projects Version 4	March 2008
Rail	Project Evaluation Manual of Rail Projects	July 2012
Road	Cost-benefit Analysis Manual	November 2010
Seaport	Cost-effectiveness Analysis Manual of Port Development Projects	June 2011

In addition to the CBA Manuals, MLIT also publishes the "Technical Guideline of Cost-benefit Analysis for Public Project Evaluation" ("TG" hereafter), which presents the general recommendations for all the CBA Manuals of transport projects (MLIT, 2009). The latest TG was published in June 2009.

The TG includes a wide range of external impacts including: air quality, water quality, noise, vibration, soil quality, biodiversity, etc. For estimating the environmental-related benefits, the TG recommends alternative cost method, hedonic valuation method, contingent valuation method (CVM) and travel cost method (TCM).

Transport projects generate different benefits to the society. These benefits may vary between projects, infrastructures, locations, timing and designs of the investment. Although the CBA Manuals generally follow the TG, there is no statutory requirement for them to adopt all the recommendations included in the TG. Some of TG's recommendations may be over-ridden or amended if they are not applicable to or appropriate for the specific infrastructure evaluation. For example, the treatment of CO_2 impacts differs between the CBA Manuals.

As the potential reductions in travel time and cost resulting from a project are relatively easy to establish and value in monetary terms, they are covered by all the CBA Manuals. Environmental-related benefits are covered only by the airport and rail CBA Manuals. The airport CBA Manual includes the procedure for estimating noise effects while the rail CBA Manual include the procedures for estimating NOx, noise and CO_2 impacts. In terms of valuation methods, these CBA Manuals recommend abatement cost approach in addition to those recommended by the TG.

It is unclear why the road and seaport CBA Manuals do not include guidance on and value of environmental impacts. Two possible reasons for the two CBA Manuals to exclude the CO_2 impacts could be related to the difficulty in determining the quantum of CO_2 emission reductions and a potential disagreement with the TG's recommended carbon value.

Discount rate and carbon value in Japan

Evaluation period and discount rate

The TG requires reporting of three set of summary results from a project evaluation. These are: Net Present Value (NPV), Cost-benefit Ratio (CBR), and Economic Internal Rate of Return (EIRR). The prescribed evaluation period for obtaining the three summary results varies with project types: 50 years for airports, 30 or 50 years for rail, 50 years for roads, and 20 to 50 years for seaports.

The TG considers two potential approaches to setting the social discount rate (SDR): the capital cost approach and the social time preference approach. It concluded that the social time preference approach is too difficult to establish in practice and therefore the TG recommends the capital cost approach.

Using the long term government bond rates (Table A.2) and the capital cost approach, the TG recommends setting the SDR at 4%, which is constant throughout the evaluation period. However, the TG does not give any reason for the constant SDR.

Table A.2. **Japanese Government Bond Yield from 1980s to 2000s**

Period	JGB 10 year Yield (Nominal, average)	JGB 10 year Yield (Real, average)
1991 to 1995	4.09%	3.91%
1986 to 1995	4.78%	3.85%
1993 to 2002	2.23%	3.10%
1983 to 2002	3.95%	3.52%

Source: MLIT (2009).

CO2 value

A “Working Group of Government Committee on Project Evaluation Method” (“WG” hereafter) was set up under MLIT in October 2008 to review the project evaluation method and to facilitate the revision of the TG. Following the growing concerns around global warming in recent years, one of the review items was the valuation of environmental impacts.

In preparation for the review process, an extensive consultation process was carried out during January 2007 to June 2008. During that process, various methodological issues were discussed with eight domestic experts from economics, civil engineering and transport research fields.

As the external impacts are typically non-market goods for which no market price is set, there are some technical difficulties in determining their values. The study team and the WG examined three potential methods for valuing CO_2 emissions: damage cost approach, abatement cost approach and emissions trading price approach.

The study team reviewed an extensive number of existing literature as well as current practices in international jurisdictions through desktop research (such as IPCC reports and academic research reports), interviews and survey questionnaire. The countries reviewed include Austria, Denmark, Finland, France, Germany, Netherland, New Zealand, Sweden, Switzerland, the UK and the US.

The key conclusions made by the study team and the WG are summarised as follows:

- “The assumptions of abatement technologies and expected damages significantly influence the values.”
- “Emissions trading price highly depends on the market design and regulations. The emissions trading market has not been well matured. The market price could be seriously biased.”
- “Abatement cost approach should be excluded because abatement technologies cannot be clearly identified. The willingness-to-pay for reducing CO_2 should be reflected into the value.”
- “National value of CO_2 emissions may have some strong message to the public and international/domestic market.”

- "Simple reviews of past studies in other countries may be biased by currency exchange rate."
- "The accuracy of estimating future damage cost may be quite low, thus the sensitivity analysis should be carried out in its application."
- "Results including state-of-the-art studies should be used. The regular updating process is strongly recommended."

Table A.3 summarises the comparisons of the three approaches to value CO_2 emissions. After considering the pros and cons of the three approaches, the WG recommended adoption of the damage cost approach using the methods shown in Tol (1999 and 2005) and the IPCC Second Assessment Report (1996).

Table A.3. **Summaries of three approaches to valuing CO_2 emissions**

Method	Advantage	Disadvantage
Damage cost approach	Easy to integrate findings of past research with increased use of meta-analysis on damage cost estimates.	Estimated values vary among studies depending on models used and assumptions around future damages
Abatement cost approach	Possible to establish a value that is consistent with government's target of CO_2 reduction in the future	Estimated values highly depend on the government's target of CO_2 reduction and the technology development in the future.
Emissions trading price approach	Theoretically reasonable as the market price	Emissions trading market has not been well developed, thus the trading price may not reflect the marginal cost.

Source: Document prepared for the Study Team meeting held in July 2007.

The review concluded that the value of CO_2 emissions should be set at 10 600 JPY per tonne of carbon[2] (or t-C) (2006 value). This value was established by first converting Tol's 1999 value (60 USD/t-C) to 2006 value using GDP deflator for the US. The resulting value was converted into JPY (2006 value) using Purchasing Power Parities (PPPs) for 2006. The review also recommends a sensitivity range based on 50% (low) and 200% (high) of the recommended value.

The recommendations made by the WG were sent to the Government Committee on Project Evaluation Method in March 2009. As a result of the discussions with the Government Committee, the TG was updated accordingly.

Discussion and conclusion

This annex summarises the current CBA practice for evaluating transport infrastructure investment in Japan. It also discusses the valuation of the CO_2 emissions in the CBA of transport projects.

Although the value of CO_2 emissions is recommended in the TG, it has not been fully introduced into all the four CBA Manuals (airport, rail, road and seaport infrastructure). This may reflect the highly decentralised system within the Japanese government. Carbon emissions typically account for only a small percentage (between 0.1%[3] and 3%) of the total transport benefits to be expected from a transport project. Therefore, there is a lack of incentive within the Japanese government to investigate into whether a common CO_2 value should be used for all modes.

Notes

1. According to Cabinet Office of the Government of Japan (CAO, 2007), over 90% of respondents in the regular opinion survey are either "concerned" or "concerned to some extent" with the global environmental issues. Since then, this percentage has been gradually increasing.

2. A tonne of carbon can be converted to a tonne of CO_2 using a conversion factor of 3.67.

3. For example, in an urban rail development project, the benefit of reducing CO_2 emissions is less than 0.1% of the estimated total users benefit.

References

Cabinet Office, the Government of Japan (2007) "Opinion poll report about countermeasures against global environmental issues," available at: http://www8.cao.go.jp/survey/h19/h19-globalwarming/index.html, accessed March 31, 2014 (in Japanese).

Civil Aviation Bureau, Ministry of Land, Infrastructure, Transport and Tourism (MLIT) (2008) "Cost-effectiveness analysis manual of airport development projects," Version 4 (in Japanese).

Ministry of Land, Infrastructure, Transport and Tourism (MLIT) (2012) Mid-term Report on Countermeasures against middle-term Global Warming taken by Ministry of Land, Infrastructure, Transport and Tourism, available at: http://www.mlit.go.jp/common/000207854.pdf,

Ministry of the Environment (MOE) (2013) "National greenhouse gas inventory report of Japan," April.

Ministry of Land, Infrastructure, Transport and Tourism (MLIT) (2009) "Technical guideline of cost-benefit analysis for public project evaluation" (in Japanese).

Ports and Harbours Bureau, Ministry of Land, Infrastructure, Transport and Tourism (MLIT) (2011) "Cost-effectiveness analysis manual of port development projects" (in Japanese).

Railway Bureau, Ministry of Land, Infrastructure, Transport and Tourism (MLIT) (2012) "Cost-effectiveness analysis manual of rail projects" (in Japanese).

Road Bureau, Ministry of Land, Infrastructure, Transport and Tourism (MLIT) (2010) "Cost-benefit analysis manual" (in Japanese).

Tol, R.S.J. (1999) "The marginal costs of greenhouse gas emissions," *The Energy Journal*, Vol. 20(1), pp.61-81.

Tol, R.S.J. (2005) "The marginal damage costs of carbon dioxide emissions: An assessment of the uncertainties," *Energy Policy*, Vol. 33(16), pp.2064-2074.

Annex B

Carbon value and discount rates in the Netherlands

Perspective of the PBL Netherlands Environmental Assessment Agency

Hans Nijland (hans.nijland@pbl.nl)

January 2015

Introduction

This annex describes the current Dutch practice of incorporating the cost of carbon dioxide (CO_2) emissions in Cost-Benefit Analyses (CBAs) for the transport sector. It gives a brief overview of the Netherland's CBA guidelines and discusses the value of CO_2 and discount rates used in the CBA process.

Incorporating climate effects in cost benefit analysis

Background of cost benefit analysis in the Netherlands

The use of CBAs in infrastructure planning in the Netherlands dates back to 1953, when Nobel Prize winner Jan Tinbergen carried out a CBA of the Dutch Delta works (Smits et al., 2006). The importance of the non-monetised effects (human lives and ecological values) was already recognised at that time. Since then, CBAs have been carried out to estimate the impacts of infrastructure projects using various methodologies. In 1998, the Dutch Ministry of Transport, Public Works and Water Management commissioned a research programme (called OEI) to develop uniform guidelines for evaluating the economic effects of transport infrastructure.

Since 2000, national projects on transport infrastructure and spatial development financed by the Ministry of Infrastructure and the Environment are required to carry out a CBA in accordance with the OEI-guidelines (Eijgenraam et al., 2000) (Dutch House of Representatives, 2000). CBAs are also required for large regional projects for which government funding is requested (Rijkswaterstaat, 2012).

The OEI-guidelines require all impacts to be expected from the planned project to be described, quantified and monetised, wherever possible. The guidelines also stipulate the discount rates and the unit value to use for different impacts (e.g. value of time and external effects). The benefit values are regularly updated by the Ministry of Infrastructure and Environment (Witteveen and Bos, 2011).

In addition, the OEI-guidelines require the effects from any project to compare a counterfactual, preferably using at least two scenarios to reflect the level of uncertainty involved, with that of a business-as-usual scenario. For infrastructure projects, the time horizon is 100 years[1] (Rijkswaterstaat 2012).

In 2013, the CBA requirement was further expanded to include non-transport sectors (such as health care, education and energy) in which CBAs were not as common at that time (Romijn and Renes, 2013). This practice aims to improve 'evidence-based' policy making.

Discount rate

The discount rate has two components – a fixed risk-free rate and a risk premium. In the Netherlands, the current fixed risk-free rate used, derived from the interest rate on the capital market, is 2.5%. As future costs and benefits are relatively uncertain, an additional project risk premium rate of 3% is added to capture the uncertain future outcomes. The overall discount rate in most cases is therefore 5.5% (2.5% + 3%). This applies for all sectors, not only the transport sector.

However, for external effects that are irreversible – such as climate change - the additional risk premium rate is 1.5% instead of 3 %. Therefore, the overall discount rate for climate change impacts is 4% (2.5% + 1.5%).

Carbon value

The impacts of CO_2 and, hence their monetary valuations, are subject to high level of uncertainty. The Netherlands used the abatement cost approach to value CO_2 emissions because they believe the approach has a lower level of uncertainty compared to that of the social cost approach. An abatement cost of 78 EUR per tonne CO_2 are currently being used (Schroten etl al, 2014).

Annema and Koopmans (2010) analysed 37 transport and 10 spatial development related CBAs performed since 2000. They found that because climate costs are relatively easy to calculate with the prescribed CO_2 value, the effects of CO_2 emission are mostly included and monetised in transport CBAs despite the estimated impacts were relatively minor. However, the climate impacts of spatial development projects were not included in any of the CBAs reviewed.

Table B.1. **Inclusion of different external effects in CBAs of transport and spatial development**

Environmental effect	Project type	Noise and emissions (CO_2, NOx, PM10)	Nature and landscape
Transport (37)		35, often monetised	18, seldom monetised
Spatial development (10)		0	8, monetised in 6

Source: Annema and Koopmans (2010).

CBAs of non-infrastructure policies

In contrast to infrastructure projects, for which the use of CBA in the decision-making process is common, obligatory and supported by well documented guidelines, the application of CBAs in non-infrastructure transport projects is less common. In the 2000- 2012 period, around 10% of all Dutch transportation-related CBAs concern non-infrastructure projects.

Examples of policy projects where a CBA is available include the policy to reduce congestion by rewarding commuters not driving during peak hours (Goudappel Coffeng, 2012), the introduction of personal mobility budgets (Annema and Van Wee, 2012) and road pricing (Besseling et al., 2005). In all three CBA studies, the impacts on climate were calculated and monetised, more or less in accordance with the OEI-guidelines for infrastructure projects.

To the best of the author's knowledge, in the Netherlands no CBA has been conducted specifically aimed at assessing climate policies within the transport sector (such as tightening of CO_2 emission standards or policies aimed at supporting the introduction of electric cars).

Conclusion

In the Netherlands, the use of CBAs in the transport sector is widespread and supported by well documented guidelines. CBAs are mostly used in the decision-making process on transport infrastructure. Within the transport sector, no CBAs exist on transport climate policies specifically.

Impacts on climate are usually incorporated in transport infrastructure CBA. Cost estimate for CO_2 is based on the abatement cost approach and is updated regularly. In the Netherlands, the long term nature of climate impacts is acknowledged by the adoption of a relatively low discount rate compared to other effects and a long time horizon.

Note

1. For residential, industrial and recreational projects with a shorter life-span, a more limited scope, for example, of 50 years may be chosen.

References

Annema, J.A., Koopmans, C (2010), "*Een lastige praktijk: ervaringen met monetarisering van omgevingskwaliteit in de MKBA,*" In MJ Koetse & P Rietveld (Eds.), *Economische waardering van omgevingskwaliteit*, Den Haag: SDU, Planologie, 14, pp. 197-213 (in Dutch).

Annema, J.A., G.P.M. van Wee (2012), "*Maatschappelijke kosten-batenanalyse van 'Persoonlijk Mobiliteitsbudget'*," TUDelft, Delft (in Dutch).

Besseling P., W. Groot, R. Lebouille (2005), "*Economische analyse van verschillende vormen*" van prijsbeleid voor het wegverkeer, CPB-document no. 87, Den Haag (in Dutch).

CE (in press), "Handboek externe en infrastructuurkosten verkeer," Delft (in Dutch).

Eijgenraam, C.J.J., Koopmans, C.C., Tang, P.J.G. and Verster, A.C.P. (2000) "*Evaluatie van infrastrucuurprojecten, leidraad voor kosten-baten analyse*" (Assessment of infrastructure projects, guideline for cost benefit analysis), The Hague (in Dutch).

Goudappel Coffeng (2012) "*MKBA spitsvrij*," rapport nr. UTA216/Bqp/1560, Deventer (in Dutch).

Kuik O., L. Brander, R.S.J. Tol (2009) "Marginal abatement costs of greenhouse gas emissions: A meta-analysis," *Energy Policy*, vol. 37(4), pp. 1395-1403.

LPBL (2010) "*Maatschappelijk rendement in het sociale domein, Een handreiking voor opdrachtgevers van MKBA's*," Amsterdam (in Dutch).

Rienstra S. (2008) "*De rol van kosten-batenanalyse in de besluitvorming*," Kennis Instituut voor Mobiliteitsbeleid, Den Haag (in Dutch).

Rijkswaterstaat (2012) "*KBA bij MIRT-verkenningen, Kader voor het invullen van de OEI-formats*," Den Haag (in Dutch).

Romijn G., Renes G. (2013) "*Algemene leidraad voor maat schappelijke kosten-batenanalyse*," CPB/PBL (in Dutch).

Rouwendal, J. and Rietveld, P. (2000). *Welvaartsaspecten bij de evaluatie van infrastructuurprojecten* (welfare aspects when evaluating infrastructure projects), The Hague (in Dutch).

Schroten, A., van Essen, H., Aarnink, S., Verhoef, E. And Knockaert, J. (2014), "*Externe en infrastructuurr – kosten van verkeer: Een overzicht voor Nederland, in 2010*", Delft, CE Delft, June (in Dutch).

Smits, A.J.M., Nienhuis, P.H. and Saeijs, H.L.F. (2006) "Changing estuaries, changing views," *Hydrobiologia*, vol. 565(1), pp. 339-355.

Tweede Kamer (2000) "*kamerstuk*" 26428 nr. 21, The Hague (in Dutch).

Witteveen Bos (2011) "*MKBA-kengetallen voor omgevingskwaliteiten: aanvulling en actualisering,*" RW1891-1-1/abdm/007 definitief 2011, Den Haag (in Dutch).

Annex C

Carbon value and discount rates in Germany

Perspective of the German Federal Environment Agency

Dr. Björn Bünger and Kilian Frey

December 2014

Introduction

Having a technically sound basis of information for estimating environmental costs is important from an environmental policy making perspective as it allows an objective debate on the costs and benefits of environmental interventions. Sound environmental cost estimates make it possible to quantify and compare the economic benefits of various mitigation measures that have impact on the health and environment of current and future generations.

Scientific advances have improved the estimation of climate costs, through better ways of estimating cause-effect relationships, better modelling of transport emissions and further developments in the estimation of emission factors. Latest best-practice estimates of environmental costs have been calculated for a number of cost categories on the basis of the UBA's Methodological Convention (UBA 2009) and its underlying research project. In 2012, the UBA released an updated version of the methodological convention for estimating environmental costs (UBA, 2012a). This paper draws heavily on the updated methodological convention and its annexes.[1]

Transport activities have many harmful effects on people and the environment, such as effects resulting from climate change, air pollution, noise, accidents, losses of biodiversity, landscape and nature, soil and water pollution, additional problems in urban areas and problems arising from up- and down-stream processes (Figure C.1).

In order to make transport activities more sustainable, a whole gamut of measures must be implemented within as well as outside the transport sector. For example, measures to incentivise the development and uptake of technological innovation such as fuel efficiency must be complemented by measures that encourage a reduction in travel and/or modal shift. A successful transition to sustainable transport can only be achieved by using a combination of these approaches (UBA 2013).

To support the research of the working group, this paper focuses on the calculation of carbon value for cost benefit analyses for transport infrastructure projects.

Figure C.1. **Average external costs 2008 for EU-27*: passenger transport (excluding congestion)**

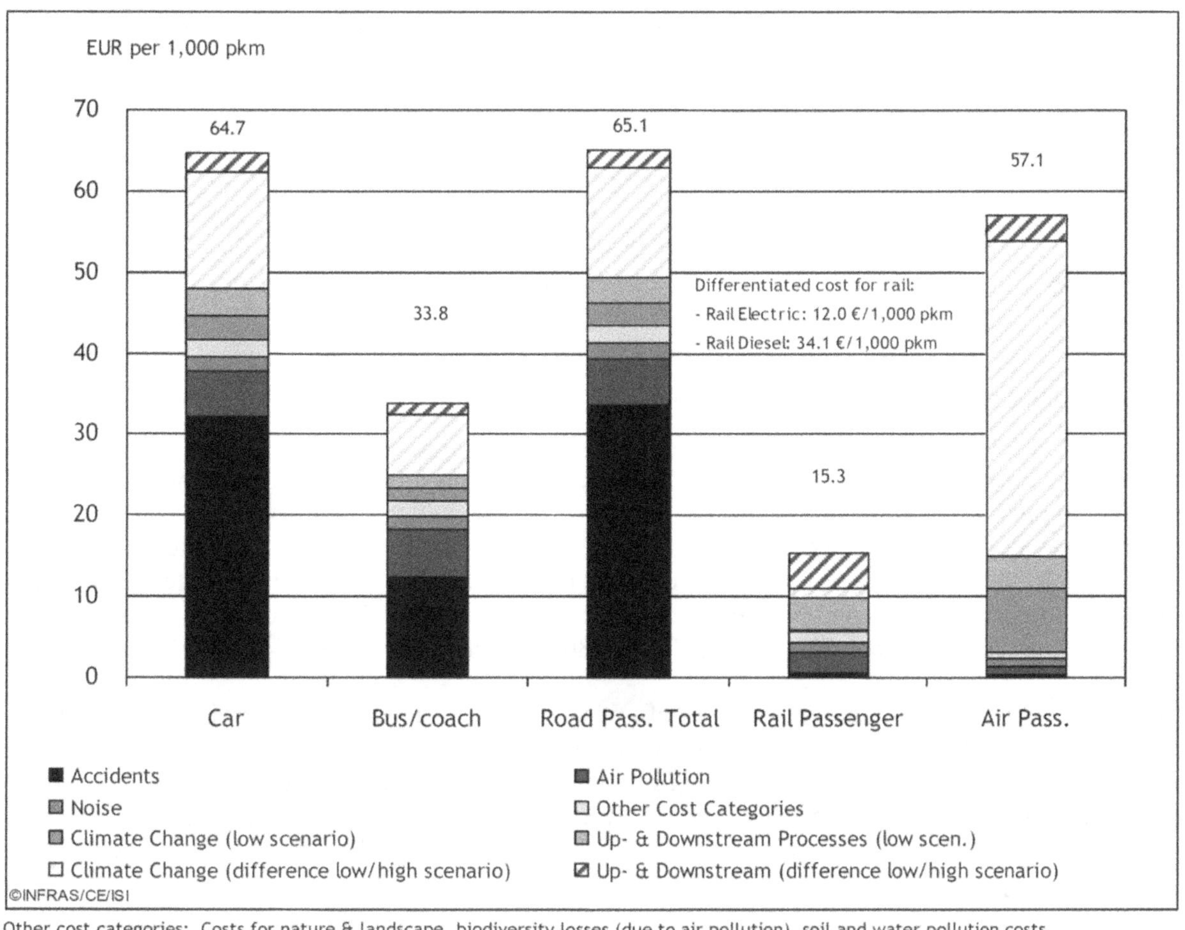

Other cost categories: Costs for nature & landscape, biodiversity losses (due to air pollution), soil and water pollution costs, additional costs in urban areas. Data do not include congestion costs.

* Data include the EU-27 with the exemption of Malta and Cyprus, but including Norway and Switzerland.

Source: CE/INFRAS/ISI (2011), External Costs of Transport in Europe, Delft.

Best practice cost rates for carbon emissions

Estimating the amount of carbon dioxide emissions (hereinafter referred as carbon emissions) and the damage caused by them in physical units and then calculating a monetary value for this damage are important for the application of cost benefit analysis. Monetisation of carbon emissions can make different costs and benefits comparable, not only between different climate mitigation policies but also between different types of public policy interventions. The benefit in the case of climate policy is the decrease in carbon emissions that can be achieved by the policy measures.

There are two broad approaches to estimate the cost of carbon emissions: abatement/avoidance costs or Social Costs of Carbon (SCC). While abatement cost estimates are a good indicator of the costs that have to be borne by an economy to achieve a specific emission target, they do not give any indication of the extent of the damage. Therefore, they are not suitable for use in cost-benefit analyses. For carrying out cost benefit analyses, estimates of the damage costs of carbon emissions (also referred as social costs of carbon) are required.

Table C.1. **UBA recommendation for best practice climate cost rates (€2010 / t CO_2)**

	Short term 2010	Medium term 2030	Long term 2050
Minimum value	40	70	130
Central value	80	145	260
Maximum value	120	215	390

Source: German Federal Environment Agency (UBA 2012b).

To estimate the climate damage costs of CO_2, the research project that led to UBA's Methodological Convention 2.0 reviewed and evaluated several studies on the estimation of damage costs. The results of the research project showed that the cost rate suggested by the Methodological Convention 1.0 (UBA 2009), of70 EUR 2000 (2000 value) per tonne of CO_2, is still valid in order of magnitude.[2] Based on a range of existing cost estimates and following the principle of erring on the conservative side, the UBA considers a best-practice cost rate should be updated to 80 EUR 2010 per tonne of CO_2.[3]

When applying the cost rates in a cost benefit analysis, UBA recommends carrying out sensitivity analyses using 40 EUR to 120 EUR per tonne of CO_2 as a reasonable range. Furthermore, UBA recommends the use of short, medium and long-term cost rates to account for the tendency for the damage costs and the avoidance costs to increase over time. For example, UBA recommends that the central cost rate for emissions in 2030 is 145 EUR 2010 and 260 EUR 2010 per tonne of CO_2 emitted in 2050 (Table C.1).

The cost rates for the greenhouse gases methane (CH_4) and NOx are calculated in the same way as the global warming potential. The costs for CH_4 are estimated at 25 times the rate for CO_2, and the costs for N_2O are 298 times the rate for CO_2.[4]

In Germany, Greenhouse gas emissions from air transport are treated as a special case with the cost rates multiplied by an emission weighting factor of two. This treatment considers the greater damage potential from high-altitude emissions.

Why discounting is necessary

In economic analyses, it is important to consider how and when costs and benefits of today's decisions will materialise over time. Benefits or costs arise today are not directly comparable to those that will arise in twenty or a hundred years' time. Discounting can make costs and benefits that arise in different points in time comparable.

Frequently, the valuation of environmental costs requires understanding the damages to be incurred far into the future: Depending on the type of effects considered (e.g. climate change), the resulting damage may only occur far, e.g. hundreds of years, into the future. The farther the damage to be assessed extends into the future, the greater the influence the discount rate has on the estimated result.

The choice of discount rate for use in evaluating inter-generational impacts cannot be substantiated scientifically because choosing a discount rate requires making certain value judgements.[5] Setting the discount rate requires, firstly, an assumption about the preferences of future generations. Secondly, it requires establishing how costs and benefits for future generations should be weighed against those for the present generation. Thirdly, it requires assumptions for determining the other parameters required for determining the social discount rate (e.g. growth rate of consumption and marginal utility of consumption used in the standard Ramsey formula).

The UBA recommends a social discount rate of 1.5 per cent to be applied as a standard value for inter-generational assessments.[6] This standard value reflects an optimistic value judgement on future economic growth potential and on changes in the marginal utility of consumption.[7] If there are any reasons to suggest a different economic growth rate or elasticity of marginal utility of consumption, the social discount rate can be adjusted accordingly.

In the context of climate effects, the best-practice cost rates recommended by the UBA are based on literature that uses a constant discount rate of 1% p.a. This corresponds to a more conservative estimate of an economic growth rate of 1% p.a. over the next 100 years. A constant discount rate applies over the entire period of time considered.

How to cope with uncertainty

There are different ways uncertainty comes into play when estimating the effects of carbon emissions in the transport sector. A key aspect of climate effects related uncertainty is the timing of the climate effects. For one thing, there is uncertainty about future technological development and how that may impact on carbon emission as well as mitigation opportunities. Secondly, as climate effects vary with the level of carbon concentration and mitigation interventions, the value of carbon is not constant over time in real terms. Thirdly, it is uncertain whether and how the relationship between carbon emissions (e.g. from transport use) and economic growth may change over time. This aspect is important for transport decision-making as transport activities account for a sizeable portion (around one-quarter) of the total amount of emitted greenhouse gases. Uncertainty regarding global economic development and its connection to carbon emissions has a major impact on the estimated damage costs of carbon.

Conventional economic valuation is based on the assumption that the effects of the alternative actions are sufficiently well known with regard to both its extent and its frequency of occurrence (or the distribution of the probabilities of occurrence). Multiplication of the extent of damage by the probability of occurrence gives the expected value of the damage. The use of expected values makes it possible to compare different courses of actions and weigh them against each other, provided that they can be converted to a uniform yardstick (e.g. monetary value). In this case, accepting the risk of a specific course of action may be referred to as accepting a “calculable risk“.

Calculating the expected values is a common practice in economic analysis whenever there are different possible states with different possible outcomes. This process enables comparison of measures that may result in different set of outcomes with different distributions of probability of occurrence. This approach, however, is based on the assumption of risk neutrality. Under the risk neutrality assumption, the product of the probability of occurrence and the effect (e.g. the extent of climate damage) has the same weighting whether a low probability is associated with a high impact outcome (e.g. climate damage) or a high probability with a low impact outcome. In reality, the expected value of a high impact outcome is often rated higher than that of the low impact outcome. This reflects the so-called risk aversion behaviour.

In principle, the expected value of the climate damage should be used for the valuation of environmental costs. If the estimated probability is based on a probability distribution, the standard deviation of the climate cost estimates should also be established.

With risk aversion, the actual value of climate effects is likely to be higher than the expected value. Therefore, the expected value should be treated as the lower limit. The risk and the reasons for risk aversion existing among the population should be described qualitatively in any economic analysis that requires the use of carbon value.[8]

Notes

1. A detailed description of the recommendations on "best-practice cost rates" can be found in "Annex B to the Methodological Convention" (UBA 2012b). In each case, the basic data and the assumptions used are documented in detail to provide a transparent picture of how the cost rates were derived.

2. When correcting for inflation, a figure of 80 EUR 2010 (2010 value) per tonne of CO_2 emitted in 2010 results.

3. This is based on a 1 percent discount rate and an equity weighting of damage by income. For details, see Annex B to the Methodological Convention, UBA (2012b).

4. Cf. IPCC (2007) and Blasing (2012).

5. In order to understand the influence of value judgements on the level of the discount rate, it is helpful to interpret the discount rate as the result of an inter-temporal utility calculation.

6. In the case of intergenerational valuations, we recommend performing a sensitivity analysis using a discount rate of zero. This case reflects a value judgement characterised by risk aversion and precautionary orientation.

7. This implies for example that the utilization of resources will improve owing to technological progress, and future generations will therefore need relatively fewer resources to satisfy their needs than the generations living today.

8. In addition, sensitivity calculations should be performed to take account of risk aversion, cf. UBA (2012a).

References

Blasing (2012) "Recent greenhouse gas concentrations," February, DOI: 10.3334/CDIAC/atg.032; http://cdiac.ornl.gov/pns/current_ghg.html.

CE/INFRAS/ISI (2011) "External costs of transport in Europe," Delft.

IPCC (2007) "Climate Change 2007: the physical science basis, summary for policy makers," Contribution of Working Group I to the Fourth Assessment Report of the IPCC.

UBA (2009) "Economic valuation of environmental damage – methodological convention for estimates of environmental externalities," Dessau-Roßlau.

UBA (2012a) "Economic valuation of environmental damage - methodological convention 2.0 for estimates of environmental costs," Dessau-Roßlau.

UBA (2012b) "Best practice cost rates for air pollutants, transport, power generation and heat generation," Annex B to "Economic Valuation of Environmental Damage – Methodological Convention 2.0 for Estimates of Environmental Costs," Dessau-Roßlau.

UBA (2013) "Germany 2050 - a greenhouse gas-neutral country," Dessau-Roßlau.

ORGANISATION FOR ECONOMIC CO-OPERATION AND DEVELOPMENT

The OECD is a unique forum where governments work together to address the economic, social and environmental challenges of globalisation. The OECD is also at the forefront of efforts to understand and to help governments respond to new developments and concerns, such as corporate governance, the information economy and the challenges of an ageing population. The Organisation provides a setting where governments can compare policy experiences, seek answers to common problems, identify good practice and work to co-ordinate domestic and international policies.

The OECD member countries are: Australia, Austria, Belgium, Canada, Chile, the Czech Republic, Denmark, Estonia, Finland, France, Germany, Greece, Hungary, Iceland, Ireland, Israel, Italy, Japan, Korea, Luxembourg, Mexico, the Netherlands, New Zealand, Norway, Poland, Portugal, the Slovak Republic, Slovenia, Spain, Sweden, Switzerland, Turkey, the United Kingdom and the United States. The European Union takes part in the work of the OECD.

OECD Publishing disseminates widely the results of the Organisation's statistics gathering and research on economic, social and environmental issues, as well as the conventions, guidelines and standards agreed by its members.

OECD PUBLISHING, 2, rue André-Pascal, 75775 PARIS CEDEX 16
(74 2015 01 1 P) ISBN 978-92-82-10791-1 – 2015

www.ingramcontent.com/pod-product-compliance
Lightning Source LLC
LaVergne TN
LVHW061252100826
845148LV00008B/1103

* 9 7 8 9 2 8 2 1 0 7 9 1 1 *